I0661080

James Mccosh

Psychology, the cognitive powers

James McCosh

Psychology, the cognitive powers

ISBN/EAN: 9783744722971

Printed in Europe, USA, Canada, Australia, Japan

Cover: Foto ©berggeist007 / pixelio.de

More available books at **www.hansebooks.com**

PSYCHOLOGY

THE COGNITIVE POWERS

BY

JAMES McCOSH, D. D., LL. D., Litt. D.

PRESIDENT OF PRINCETON COLLEGE, AUTHOR OF " INTUITIONS
OF THE MIND," " LAWS OF DISCURSIVE THOUGHT,"
" EMOTIONS," " PHILOSOPHIC SERIES," ETC.

NEW YORK
CHARLES SCRIBNER'S SONS
1889

PREFACE.

FOR the last thirty-four years I have been teaching Psychology by written lectures to students in Ireland and America. From year to year I have been improving my course, and I claim to have advanced with the times. As Uncle Toby's stockings were so often darned that he was not sure whether there remained a single thread of the original fabric, so my prelections have been so constantly mended that I do not know that a single sentence remains of my early lectures.

I certainly wish this little work to be used as a text-book, and would thus widen and prolong my teaching power. But people say "dull as a text-book." In physical science and in literature they illuminate their books (as in the old missals) by figures. We cannot do this in mental science, as our thoughts have not forms nor colors. I maintain, however, that they have livelier features. I have sought to avoid dryness by illustrating mental laws by examples taken from human nature. As general laws are drawn from particular cases, so they are best understood by concrete facts coming under our experience.

It will be shown in this work that the honest and care-

ful study of the human mind in an inductive manner un-
dermines the prevailing philosophic errors of this age ;
saves us from Idealism on the one hand and Agnosticism
on the other; and conducts us to Realism, which in a
rude state was the first philosophy, and when its ex-
crescences are pruned off will be the last.

Following the example set by several distinguished
writers, I have carried out my exposition of the faculties
by instructions as to their improvement.

I hope to add to this little work another on the Motive
Powers of the Mind, including the Conscience, Emotions,
and Will. I have already so far anticipated this by my
work on the Emotions.

I have to express my obligations to my former pupils :
to Professor Macloskie for diagrams, and to Dr. Starr
and Mr. J. M. Baldwin for the exposition of certain
points which they have studied carefully.

PRINCETON COLLEGE, *June*, 1886.

CONTENTS.

INTRODUCTION.

BOOK FIRST.

CHAPTER I.

CHAPTER II.

BOOK SECOND.

CHAPTER I.

CHAPTER II.

CHAPTER III.

CHAPTER IV.

CHAPTER V.

THE COGNITIVE· POWERS.

INTRODUCTION.

DEFINITION OF PSYCHOLOGY. — METHOD OF INVESTIGATION.

PSYCHOLOGY is the science of the soul. The word is from *psyche*, soul, and *logos*, speech or reason. By soul is meant that self of which every one is conscious. Science is systematized knowledge, and when we arrange the knowledge which we can acquire of the soul, we have the Science of Psychology.

In constructing it we proceed on the Method of IN-DUCTION. This is distinguished from Deduction, in which, as for example in mathematics, we proceed from assumed or admitted principles to truths derived from them. In Induction we gather in (*induco*) facts, but always with a view of discovering an order among them and arranging them. It is found that in all nature physical and mental facts proceed uniformly or regularly, that is, according to laws. This is the case in physics: matter attracts matter inversely according to the square of the distance. It is also so in psychology: like tends to recall like. It thus comes to be the end of science to discover laws. Psychology may be more fully defined as that science which inquires into the operations of the conscious self with the view of discovering laws.

Induction begins with OBSERVATION. In botany we collect plants and look at their forms and habits. In

1

psychology we notice mind as it operates and mark its various states. In Induction we also employ EXPERIMENT, which is a mode of observation in which we artificially place the agents of nature in new circumstances that we may perceive their action more distinctly: thus, in order to determine whether all bodies fall to the ground at the same time, we put a guinea and a feather in the exhausted receiver of an air-pump, that we may note the time they take. to descend, independent of the resistance of the air. In like manner, in studying the human mind we place objects before it that we may find how it is affected by them: thus, in order to determine how the conscience acts, we direct it to a cruel or a beneficent act; and how the emotions are raised, we call up objects fitted to gratify or disappoint our springs of action.

Both in physical and in psychical science we begin with and proceed throughout by Observation Proper and Experiment. But there is a difference in the agent or instrument of observation in the two departments. In the former we employ the senses, such as sight and touch, aided by such instruments as the telescope, microscope, and blow-pipe, and we weigh and measure the bodies. In psychology we make our observations by Self-Consciousness, which is the power by which we take cognizance of self as acting, say as thinking or feeling, as remembering the past or anticipating the future, as loving, fearing, resolving.

Self-Consciousness may give us information directly or indirectly. (1.) We may notice the states of the soul as they flow on, our judgments and our fancies, our joys and our griefs. (2.) By a brief memory we may throw back our mind on the past and recall what has been under the consciousness in a given time; say during the past hour, when we were earnestly thinking, or under deep sorrow, or cherishing ardent hope. (3.) We may gather what has passed through the minds of other people from their words or their deeds : as we listen to them, as we read their writings — say biographies or histories, poems or novels ; or as we observe their conduct in ordinary or in trying circumstances. We understand what these are because of our own conscious experience. Our field of view is thus enlarged indefinitely, and becomes as wide and varied as our intercourse with mankind and our reading.

It is proper to add that light may be thrown on the operations of the mind by the physiology of the brain and nerves. We know objects external to the mind by the senses, and it is important that we know how the senses work. We are not to suppose that the brain and nerves think ; but still the rise and even the nature of certain mental affections depend much on these, and light may be thrown on the action of the conscious soul by a careful study of the parts of the body most intimately connected with the action of mind. Observation in psychology is to be conducted mainly by self-consciousness, but may be aided by the physiology of the cerebro-spinal mass.

Beginning with the observation of states or affections of mind, we then note their resemblances, differences, and other relations, and can thus coördinate them, place under one head those that are like, and give them a name by which we can speak of them. Thus we find that in certain exercises we notice the external objects before us, and we give to them the common name of sense perception ; that in others we recall the past, and this we call memory ; or we picture unreal objects, such as a mermaid, and this we designate imagination ; or we infer from what is given or allowed something else implied in

It, and this is said to be reasoning ; or we distinguish between good and evil, and this we speak of as conscience ;
or we are affected with sadness, which is emotion ; or we
resolve to do a certain act, which is will.

According to this view Psychology should have as its
province the operations of the conscious self, leaving to
Physiology the structure of the organism.   These two,
the soul and the organism, have mutual connections, and
the sciences which deal with them may throw light on
each other, but all the while they are to be carefully dis
tinguished.

All parts of the organism fall under the science o
physiology and not of psychology.   But were it only to
enable us to distinguish between physical and psychical
action, it is necessary to look at certain actions of the
nervous system most intimately connected with mental
action.   All along the spinal column there is automatic
action which is reflex.   There is a cell called a ganglion,
into which one nerve enters and from which another
goes out.   On the former being stimulated at the extremity, an action passes along to the centre, and then
motion proceeds along the latter.   We have an example
in the frog's leg moving when it is pricked.   Here
there is neither sensation nor volition.   No sensation is
felt till the action goes up to the brain.

The central mass of the brain consists of " basal gang-
lia " (the *optic thalami* and *corpora striata*, as in Fig.
2), from which commissures of white fibres radiate to the
gray cortical matter.   The gray matter, which is at the
surface, is cellular, and is most intimately connected with
mental action ; the white matter is in the deeper parts,
and consists of masses of fibres running in different directions, which are supposed to be mainly transmissive.
The communication from the spinal cord is up by the

medulla oblongata (Fig. 1, Fig. 2 L.) and the crura cerebri to the corpora striata and optic thalami; and in all the
higher animals there is a large transverse bridge called
corpus callosum, which connects both sides of the brain.

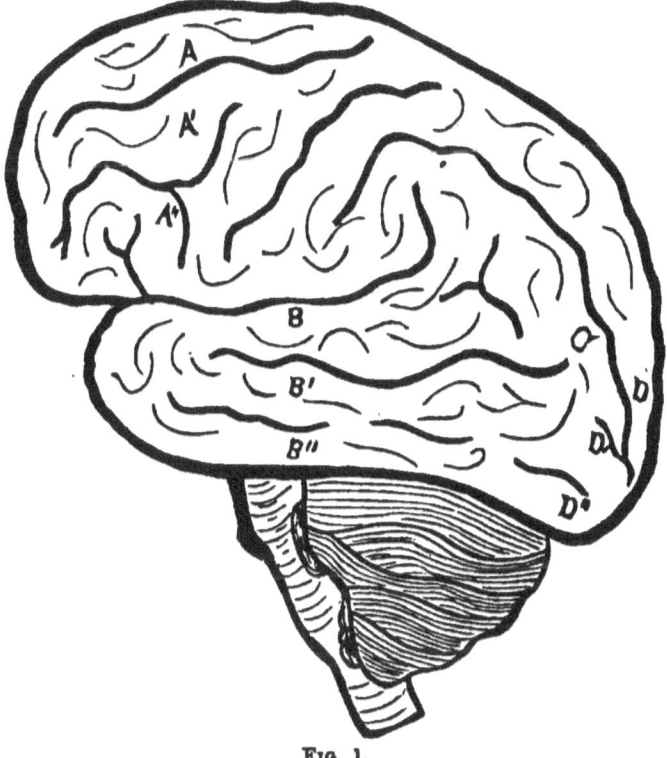

FIG. 1.

BRAIN, external view, showing *cerebrum* above, *cerebellum* and *medulla oblongata* be-
low and behind; front of brain to your left side. A, A', A'', the frontal lobes;
B, B', B'', the temporo-sphenoidal lobes; C, the angular gyrus (seat of vision);
D, D', D,'' the occipital lobes.

The nerves which carry the action to the brain are
called afferent, those which carry out the action from
the brain are efferent. The former are Sensor, the latter
Motor. The former are denoted by P S, the latter by

A M, as the former are posterior and the others anterior in the human frame.   There is also a sensori-
motor system, of which sneezing is an exercise, in which
there is sensation and motion but no volition.   The
action along the nerves occupies time which has been

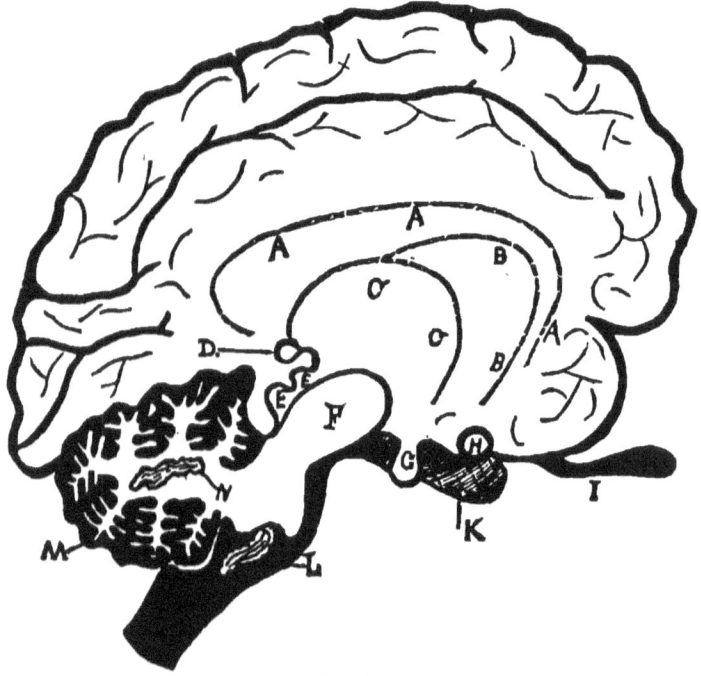

FIG. 2.

THE BRAIN, median vertical section: front of brain to your *right* hand. A, A, *corpus
callosum*; B, B, *corpora striata* (laterally from median plane); C, C, *thalami optici*;
D, pineal gland (deemed by Descartes the seat of the soul); E, E, *corpora quad-
rigemina*; F, the *crura cerebri*; G, the pituitary body; H, the commissure of
the optic nerves; I, the olfactory lobe; K, temporal lobe; L, *medulla oblongata*;
M, cerebellum, with (N) its axial part, and the *arbor vitæ*.

measured with approximate accuracy.   Thus, the action
to the brain travels at the rate of 140 or 150 feet in the
second.   The action from the brain travels about 100
feet in the second.   The rate is slowest in sight, next
slowest in hearing, and quickest in touch.

SECTION II.

PROOF OF THE EXISTENCE OF MIND.

But, it is asked, what evidence have we of the existence of the soul? The answer is that we know its existence intuitively, by looking in upon it as it is acting. We are conscious of it as perceiving, imagining, thinking, resolving, hoping. fearing, loving. We have thus evidence primary and not merely secondary, original and not derived ; as certain as we have for matter.

But, then, it is asserted that mind is not different from matter, that it is a mere modification of matter. It can be shown in opposition, first, that we know the two by different organs. We know matter by sense-perception ; we know mind by self-consciousness. We cannot by the senses observe any pure psychical act. We can touch our own body or our neighbor's, but we cannot touch our own soul or his. We can see a colored surface, but we cannot see a thought. We can taste food, but not an affection of love or of fear. We can hear a sound, but not a reproach of conscience. We can smell a rose, but not a feeling of beauty.

Secondly, we know mind and matter as possessing different properties. We know matter as extended, that is, as occupying space and being contained in space. We further know body as resisting our energy and acting on other bodies. We know mind, on the other hand, as apprehending, judging, reasoning, distinguishing between right and wrong, as under emotion, as wishing and resolving. It is acknowledged that we know things only by their properties, and we know mind and matter to be different by their manifesting different properties. It is a favorite position of some in the present day, that the two are correlates of one another, that they are two sides or aspects of one and the same thing. But can we attach any meaning to what we say when we describe thought as a side or aspect of a stone or of an acid or a piece of timber? Just as little can we understand or conceive that our musings, our fancies, our resolutions should have solidity, durability, elasticity, hardness, softness, porosity, pressure, gravity. We thus know them as different things and should so investigate them, and seek to determine the properties of each. We may afterwards inquire into their points of connection.

SECTION III.

CAUTIONS TO BE ATTENDED TO IN THE STUDY OF THE MIND.

(1.) *Certain ideas must be left behind.*—We must not take materialistic conceptions with us into psychology. In the natural history of the mind things without us are noticed before the things within us. We are in consequence exposed to a temptation in beginning in youth or mature age the discussion of psychical questions: we apply ideas got from matter to mind. We need to guard against this. Thus we are not to allow ourselves to look on mind itself or any of its operations as occupying space, as extended, or as having figure, as having weight or levity, height or depth, elevation or depression, attraction or repulsion, solidity or elasticity, motion or rest, light or darkness, warmth or frigidity. We have come to an entirely new country, and we must learn to accommodate ourselves to the people, to their laws and customs, and in particular we have to learn their language.

(2.) *We have to beware of the misleading influence of language derived from material objects.*—As the individ aal looks without before he pays special regard to the mind, so in the natural history of society there is an acquaintance with physical nature before there is a study of our mental nature ; and our first language is sensible rather than spiritual. So when philosophers begin to study the human mind they have either to coin and employ a new language, which is very irksome, or they have to adopt the old phrases expressive of external and extended objects. But the old idea is apt to come in covertly with the old phrase as we use it. Thus the original meaning of " idea," signifying image (first turned to a philosophic purpose by Plato), is apt to come into our minds (as it does in Locke's philosophy) with the phrase as applied even to a mental concept or notion. The terms employed in various languages to denote the mind — *psyche* in Greek, *anima*, *spiritus*, in Latin, *ruah* in Hebrew, and *âtman* in Sanskrit, originally signified breath or wind. " Feeling," at first signifying an affection of touch, now signifies an emotion such as hope and fear. " Emotion " is literally a moving out. " Impression," a fatal word introduced formally into philosophy by Hume, denotes a mark left on a soft substance by a hard. " Understanding," now denoting the intellect, refers to something standing under. " Apprehension" and " conception," applied to mental acts in which we lay hold of or bring things together, meant at first a seizing by the hand. We cannot afford, even at the present advanced stage of inquiry, to lay aside such phrases ; but when we use them we must strip them of their materialistic associations.

SECTION IV.

CLASSIFICATION OF THE FACULTIES.

As there are some who doubt whether the mind can be represented as having Faculties, or at least separate faculties, it will be necessary to lay down some explanations and limitations.

I. *The mind evidently possesses power.*

Matter itself possesses power. It is acknowledged to have properties, and what are properties but powers? It has, for example, a gravitating, a chemical, an electric power. Physical science is seeking to determine the precise law, rule, and expression of the powers of body. If matter has power, much more has mind. The powers of mind are different from those of matter. If the one has attractive and magnetic powers, the other has powers of understanding and emotion. The mind has powers, but not all possible or conceivable powers. Its powers are bounded. Thus we cannot tell what is doing at this moment in the planet Venus or the constellation Orion. Just as physics would determine the precise rule and limit of gravitation or chemical affinity, so psychology should try to ascertain and express the precise laws of such powers as the memory, the imagination, the conscience.

II. *That there are different powers in the mind is evident from the differences in the mental states and affections of different persons.*

This conclusion might be drawn from the very differences between man and brute. The lower animals possess powers common to them and human beings ; but there are others, such as the discernment of moral obligation, which are peculiar to man.   But the inference can be drawn more directly from the circumstance that one man is distinguished for powers which are either not pos-sessed by other men or possessed only in an inferior de-gree.   Thus one man has a great tendency to observe causes, another resemblances; one has keen emotional sensibility ever ready to flow out, another a resolute will. It has further to be noticed, as decisive of the whole question, that these capacities and inclinations may be-come hereditary and go down from father or mother to son or daughter.

III. *This is further evident from the circumstance that we are not always exercising every faculty or the same faculties.*

In every given state of mind there seem to be more than one power in exercise.   But all the mental powers are not in action, or at least in intense action, every in-stant.   At this moment I may be looking at the paper before me, and at the same time collecting my thoughts to write this paragraph.   Immediately after, I may be looking at the same paper, but my mind may have wan-dered off to some imaginary scene in which I and my friends are figuring.   From such a case we see that memory is different from imagination, for I was remem-bering when I was not exercising imagination, and imag-ining when I was not remembering.   It is evident, too, that both memory and imagination are different from sense; for we had the senses in the one case without memory and in the other without imagination.

Some would say that what are spoken of in these ar-ticles are not different faculties, but different modes of consciousness.   I am not sure that this is an improved statement or the correct statement.   Our perceptions, recollections, judgments, are not modes of consciousness the accurate account is that self-consciousness observes them, and they must exist in order to their being noticed. But even though they were modes of consciousness, the question would immediately arise, What are these different modes? And in answering we would be brought back to different powers leading to the diverse manifestations of consciousness.

IV. *The faculties are powers of one indivisible mind.*

They do not differ from each other, as the hand does from the foot, or the lungs from the heart. They are powers of one existence possessing a variety of attributes.

V. *The faculties are not to be regarded as necessarily operating one after another in regular order or at different times.*

The properties of matter often act simultaneously. At the same time that the iron is chemically combining with oxygen to form rust, it is attracted to the earth by gravitation, and yet we regard the gravitating and chemical powers as different. On a like principle we are constrained to regard the capacity of sense-perception, when the object is present, as different from the memory, when it is absent. It seems clear that several of the mental powers may be blended in one act. Thus at the same time that I am judging or deciding, I may be under the influence of hope or fear, of benevolence or prejudice. How many diverse powers may be exercised at one and the same time in that blade of grass, or in our finger: the gravitating, chemical, electric, vital; no one can tell how many. There may be a like number and diversity of powers at work in certain of the exercises of the mind, as when men are solving perplexing problems, speculative or practical, or rising to the higher flights of genius.

VI. *It is difficult to form a classification of the faculties which deserves to be regarded as complete.*

This arises from a variety of causes. It may proceed from human incapacity, from the difficulty of penetrating phenomena which are so fugitive — that is, so briefly under the view — and so complicated, and from the circumstance that the faculties very much run into each other. This is a hindrance not peculiar to psychology. How difficult do botanists find it to draw out an arrangement of the vegetable kingdom which may include all and exclude none, which may combine the like and separate the unlike. Yet they do contrive to draw out such a classification as is fitted to bring into view the sameness and difference of plants. We may in like manner so distribute the operations of the mind as to unfold their characteristics and their distinctions.

VII. *There may be a classification of the faculties embodying much truth and of eminent practical utility, though not professing to be perfect.*

It is true that the mind is one, but it manifests itself in a variety of ways, and its characteristic operations must be carefully noted and their peculiarities unfolded. It is only when the acts are marked, distinguished, classified, and named that we can be said to have any adequate idea of the nature of the mind. For practical ends, for the purposes of the orator, the poet, the advocate at the bar, and the preacher in the pulpit, even for ordinary letter-writing and conversation, there must be distinctions of some kind drawn as between the head and the heart, between the imagination and the judgment, be-

tween the understanding and the will. It is the business
of the psychologist to seize upon real distinctions and
unfold them as accurately as possible, and in this he cannot err to any extent, provided he follow a careful observation and be ready to confess that while he exhibits the
truth, it is not the whole truth, and that however much
we know there is always more to man unknown.

VIII. *In proceeding to distribute the powers it is first of
all desirable to have some such division as that which we
have of the physical world into the mineral, the vegetable, and the animal kingdoms.*

The Eleatic School (500 B. C.) had a loose division
of what are now called the intellectual powers into Sense-
Perception, probable Opinion (δόξα), and Reason (λόγος).
Plato had a like threefold division, and had a further
division of what is now called the Motive Powers intc
Sensual Feelings, Impulse, and Love. Aristotle gave a
better division into the Gnoetic or Gnostic, translated
Cognitive, and the Orective, translated Appetent or
Motive. This twofold division reappears in the distinction between the Understanding and the Will, the Intellectual and Active Powers, and popularly the Head and
the Heart. Of a later date some have felt it necessary
to draw distinctions of an important kind between the
various powers embraced in the Will or Heart, and this
led to a threefold division, the Cognitive, the Feelings,
and the Will, a classification adopted by Kant and Hamilton. In this division the Senses must be included
under either the Cognitive or the Feelings, or divided
between them. To avoid this awkwardness there is a
fourfold distribution, the Senses, the Intellect, the Feelings, and the Will. It should be observed that in this
distribution, the Conscience or Moral Faculty has nc

place; and those who have carefully noted its operations will acknowledge how difficult it is to bring it, with its peculiar ideas of right, wrong, and duty, under any of the heads named. To avoid these and other difficulties the following, embracing all the others, is submitted as a good provisional division, fitted to expose to view the leading attributes of the mind.

N. B. It should be noticed: (1.) The Conscience, which is both a Cognitive and a Motive Power, has the attributes of both the two heads. (2.) The Compositive Power or Imagination can be called a Cognitive Power only with the explanation that it is cognitive not as it knows existing objects, but inasmuch as its ideas are re-productions of cognitions.

FIRST GROUP, THE COGNI-
TIVE.

I. THE SIMPLE COGNITIVE OR
PRESENTATIVE.

1. Sense-Perception.
2. Self-Consciousness.

II. THE REPRODUCTIVE OR
REPRESENTATIVE.

1. Retention.
2. Recalling Power or Phantasy.
3. Associative.
4. Recognitive.
5. Compositive.
6. Symbolic.

III. THE COMPARATIVE, DIS-
COVERING RELATIONS

1. Of Identity.

2. Comprehension.
3. Resemblance.
4. Space.
5. Time.
6. Quantity.
7. Active Property.
8. Causation.

SECOND GROUP, THE MO-
TIVE.

IV. THE CONSCIENCE A COG-
NITIVE AND MOTIVE POWER.

V. THE EMOTIONS, WITH
MOTIVE PRINCIPLES.

VI. THE WILL.

1. Wish.
2. Attention.
3. Volition.

SECTION V.

EDUCATION OF THE FACULTIES

It is often said that education should proceed philosophically. But there is no agreement among those who
hold this view as to what philosophy is, some preferring
the Scottish, others the Hegelian, and a number in the
present day the Sensational or Materialistic philosophy.
It is more correct and definite to say that education
should proceed psychologically, and when it does so it
proceeds philosophically. But what does this mean?
It may mean two things somewhat different and yet connected, and both important. It may mean that we edu-
cate the faculties. This should be one of the aims, one
of the main aims, of education. Our faculties are in the
first instance mere capacities with a tendency to act.
They are in infants in the form of a seed, or germ, or
norm, and need to be cherished in order to grow and to
be useful. They are all capable of being trained and
should be trained, and education, private and public,
should undertake the work. But the statement that education should proceed psychologically may mean something more. It may signify that education should proceed according to the genesis and natural growth of
the powers. It implies that we begin with the lower
and go on to the higher powers. Our psychology, if properly constructed, may greatly aid the science of education. It shows us what the faculties are, what their laws
and modes of operation, and it is by knowing these that
we are able to train them. It should show us what
powers first appear, and how one power grows out of another; and thus lead us to discover what branches should
be taught and in what order, what should be taught to

children and what to those farther advanced. For special purposes, scientific, professional, or practical, greater pains may be taken with some of these powers than with others, but at the same time all should be so far cultivated as to keep the mind properly balanced, and to prevent it from being one-sided, exclusive, partial, and prejudiced. Now, these topics may legitimately be taken up in a work on psychology, at least in an incidental way, as we proceed.

2

BOOK FIRST.

THEY are so called because they give us knowledge in its simplest form — that is, as will be explained, in the singular and in the concrete; and because the objects are now present and presented. Other faculties are also cognitive, but they proceed on the knowledge acquired by these primary powers, and they form composite, abstract, and general notions. The other faculties also look at objects, but these, as in memory, for instance, are not present; they have been in the mind before, and are not presented, but represented. Let us try to discover what must be the first exercise of the conscious mind. It must, I apprehend, be knowledge.

Knowledge the First Mental Exercise. — By this is not meant scientific, that is, arranged knowledge, but knowledge of an object as it presents itself single and with its qualities. We may suppose that it is a knowledge of our bodily frame, say of the tongue or nostrils, or foot or finger. Not that we as yet know that it is the tongue or toe, or a member of our complex bodily frame which in its entirety may as yet be unknown; yet it may be knowledge, forming the basis of all higher knowledge, abstract and general.

(1.) Our knowledge must begin with things apprehended as singular. Out of the single things we form general notions by observing points of resemblance: as having seen a number of flowers of a particular type we form the class "rose." This knowledge is also concrete, that is, of things with qualities. This rose is known as having a certain form and color. Out of the concrete we form abstract notions, such as redness.

(2.) If the mind did not begin with knowledge, it could never reach it by any process of thought. "How can we reason but from what we know?" and if we have not knowledge in the premises, we are not entitled to put it into the conclusion. David Hume started with "impressions," as of colors, and "ideas," mere reproductions of these, such as remembered colors, and thus introduced the most formidable skepticism ever propounded. We meet the skepticism at its entrance, by holding that our first conscious experience does not consist of impressions, but is a knowledge of things.

This generic group comprises two special powers: (1) Sense-Perception, or knowledge by the senses; (2) Self-Consciousness, or a knowledge of self in its present state.

CHAPTER I.

SECTION I.

ITS NATURE: ORIGINAL, INTUITIVE, POSITIVE.

By this power we get a knowledge of things affecting us, external to ourselves and extended. The things thus known· we designate " matter," or " body," corresponding to which we have convenient adjectives, "material" and " bodily."

In perception, the mind takes cognizance of something external to the perceiving mind. The ego comes, as metaphysicians say, to know the non-ego, or, as I prefer saying, the self knows the not-self. It is not a sensation merely that is given us, or a feeling; it is not an idea or an apprehension, or a notion or a conception ; nor is it a belief or faith. It is more than a sensation or a feeling, which may accompany the perception. The experiences denoted by the other phrases come afterwards, and imply a previous knowledge. It is not the exact or full truth to say that I feel an external object, or that I have an idea of it (which I may have when it is not present), or that I apprehend it, or have a notion of it, or believe in it ; the correct expression is, that I have a knowledge of it, or that I cognize it, a phrase which gives us a corresponding adjective and noun, cognitive and cognition. It has to be added that the object is known as affecting us The primary knowledge is thus both objective and sub

jective: that is, of an object, but this as perceived by the
subjective mind. The two are together in the act of cognition, but they are after all separate, and are separated
by every intelligent mind which does not mistake the
not-self for the self, and never confounds the perception
with the object perceived. The confounding of them is
the work of bad reflective or metaphysical philosophy,
and not of spontaneous thought. Let us determine some
points as to our knowledge by the senses.

I. We have sense-perceptions which are ORIGINAL and
not derived. Were they not given us by an original endowment they could never be obtained by experience, by
inference, or any other process. Experience, properly
speaking, is only a repetition and collection of what we
have passed through, and if there be not knowledge in
the original experiences, it cannot be had by accumulating
them. As little can it be had by reasoning, except from
premises which contain knowledge of material objects ;
without this there would be an evident illicit process, that
is, we have more in the conclusion than we have in the
premises.

II. Sense-Perception is INTUITIVE, that is, we look
directly on a material object. I do not inquire at present what is the precise object perceived, whether it be
in the bodily frame or beyond it ; how far in, if it be in
the bodily frame, how far out, if it be beyond it. Explanations will require to be given and distinctions drawn
before we can determine what is the precise object. But
whether in the body or without the body, there is an object perceived directly as extended and affecting us. This
is the simplest hypothesis, and is accompanied with no
difficulties. Every other supposition lands us in inextri-
cable perplexities. It is certain our consciousness so tes-
tifying, that we do know material objects; but nothing

coming between us and the object could impart the cognition.

III. Sense-Perception is POSITIVE, and not merely Phenomenal or Relative: that is, it is of things as they appear, and not of appearances without things, of things known, and not of the relations of things themselves unknown.

This proposition is laid down in opposition to two views commonly entertained in the present day. The one is the Phenomenal theory of knowledge, which holds that all we can know originally are appearances, and that we cannot know what things are except by some further process, or that we cannot know whether there are things or no. We meet this unsatisfactory doctrine by maintaining that we cannot know appearances except as the appearances of a thing appearing. We do not know all about this thing, we may not know much about it, but we are sure that it exists when it appears to us, and that it is known to us under a certain aspect or as doing something. Even an echo, coming from a hollow in which nothing is seen, has a reality in vibrations of the air reaching the tympanum of our ear.

Closely allied to this theory is that of Relativity, according to which we do not know things, but merely the relation of one thing to another, to ourselves, or to some other things. Now this is to reverse the proper order of nature. We must so far know things before we can discover their relations. In the discovery of relations we so far know the things; we know them as having the qualities which bring them into relation. These positions are laid down in opposition to three theories which have been widely entertained, and which it may be useful to look at and examine.

SECTION II.

THEORIES OF SENSE-PERCEPTION : IDEAL, INFERENTIAL, PHE-
NOMENAL AND RELATIVE : NATURAL REALISM.

THE IDEAL THEORY. According to it the mind does not perceive the material object, but some idea or repre-. sentation of it, some medium or *tertium quid* coming between the object and the perceiving mind. This explains nothing, and brings in perplexities in addition to those which belong to the subject itself.

It was introduced to solve the difficulty supposed to arise from matter being thought to act on mind and mind on matter. The principle was laid down as early as the days of Empedocles, that like could act only on like. So it was necessary to bring in something to interpose between the object perceived and the perceiving mind. According to Democritus, the expounder of the atomic theory of matter, images (εἴδωλα) composed of the finest atoms floated from the object to the mind. Lucretius has expressed the theory in " De Rerum Natura," lib. iv. 48–53 : —

> " Dico igitur rerum effigias tenuisque figuras
> Mittier ab rebus summo de corpore rerum,
> Quoi quasi membranæ vel cortex nominitandast,
> Quod speciem ac formam similem gerit ejus imago
> Cujuscumque cluet de corpore fusa vagari."

It has appeared in one form or other ever since. It takes a grosser and a more refined shape. Some look on the idea perceived as a sort of material figure, like the image in a mirror or that formed on the retina of the eye when an object is before it. This removes no difficulty ; for if this be a material figure, how can so different a substance as mind perceive it? With most modern metaphysicians the theory has taken a more spiritual form. Some make the idea an affection of the brain. Most of its supporters do not know what to make of it. With the more sensible the idea is merely the mind apprehending the object; but in this case the idea is not the object looked at, but the mind looking at it. Locke speaks everywhere of the mind perceiving the idea rather than the thing, and has thus confused his realistic philosophy and made knowledge consist in the discovery of the conformity of our ideas to one another, and not their conformity to things. And so the question was raised and has been much discussed by Thomas Reid, Sir William Hamilton, and the Scottish School of Philosophy, as to whether it is necessary to suppose that there is anything coming between the perceiving mind and the thing perceived. To allege that there is such a middle agent is at best a hypothesis of which there can be no positive proof. As a hypothesis it explains nothing, but rather perplexes everything by bringing in agents, of the existence of which we have no proof, and which, if they did exist, would demand new explanations. For we have now to account, not for the action of body on mind, but for the action first of body on this idea, and then the action of this idea on the mind. The simplest, the most satisfactory account is that body acts on mind, and that we perceive the very thing.

THE INFERENTIAL THEORY. — According to it the knowledge of objects external and extended is got by inference from something else; from a sensation or from an undefinable thing called an impression. Some regard the argument as legitimate, and believe in the existence of body. Some look upon it as illegitimate, and so hold that there is no proof of the existence of matter.

Certain metaphysicians of the French Sensational School, such as Destutt de Tracy and Dr. Thomas

Brown of Edinburgh, held that from a sensation in the mind we argue the existence of an external world, and justified the inference. According to this theory there is first a sensation and then an inference that there is an external object, the cause of it. For example, the child notices what it afterwards learns to characterize as the face of its mother. It finds that it cannot reproduce this at pleasure, and that there is nothing within itself to produce it, and it concludes that there must be something external acting as its cause. It is supposed that an accumulation of such experiences gives us the idea which we have of matter. Now, there are manifestly many assumptions in this supposed process. First, it is assumed that everything beginning to be must have a cause. Brown regarded this principle as intuitive, and so was entitled to use the principle. It might be difficult, however, to prove that if a child were shut up within its own self it could at an early date, or at all, arrive at a belief in invariable causation merely from experience, for its experience would habitually be of events without a known cause. But granting that it could, the difficulty arises, How could the mind think or imagine anything external of which it has no experience, till, as is supposed, it has drawn the inference? But a more formidable, I believe an insuperable, objection remains. It is certain that in our natural idea of, or belief in, an external world, we regard it as extended. But how have we got this idea? From the experience of a sensation which is without extension, we are not entitled to argue the existence of an extended object, as we would have something in the conclusion, namely, extension, not in the premises. The reasoning being thus illegitimate, we are driven to one or other of two conclusions · one, by far the most reasonable, that we perceive extended objects at once and intuitively.   The other is that matter is and must be forever unknown to us, — the conclusion drawn from Brown's view by J. S. Mill and his school, which sets aside our intuitive convictions.

THE PHENOMENAL AND RELATIVE THEORIES. — Reference has already been made to these.   The former is a Kantian modification of Hume's doctrine that all the mind perceives through the (supposed) senses are impressions.   Kant saw at once that these impressions were not knowledge, and could not give knowledge. Not wishing to assume anything not allowed him by the skeptic, he took the position, Let us assume that there is nothing but appearances, and agree to call the things thus primarily before us presentations, without asserting what they are ; and then he proceeded by a series of subjective forms to fashion them into a grand intellectual system.   But as he had not objective reality in what he started from, he never could reach it by any formal process of thought.   So his philosophy commenced with appearances and culminated with subjective forms.   I meet this theory from the beginning by insisting that appearances must be appearances of something, are in fact things appearing ; and that in our first mental operations we know things presenting themselves.   According to the other and allied theory we know merely relations.   True, we are able to discover relations, but they are relations between things so far known.   Our knowledge of relations is of things real or imaginary as related. We have as clear evidence that we know things as that we know the relations of things.

NATURAL REALISM, OR IMMEDIATE PERCEPTION. — According to it we perceive the external object directly. That object may be in our frame, or in a body affecting our frame.   Upon our primitive knowledge we may

build other knowledge by further experience, and by legitimate inferences. But all our experiences throw us back
on an immediate knowledge of matter. All our reasonings about body imply a primitive cognition on which
they proceed.

It must be left an unsettled question, in regard to
which we may have to seek and obtain further and further light, what is the precise object we perceive by the
senses generally, and by each of the senses. Before the
time of Berkeley it was generally believed that we at
once know distance by the eye. Since his time it has
commonly been acknowledged that in this knowledge
there are gathered observations and reasoning involved.
But these acquired perceptions imply primary ones on
which they proceed. It is by such facts, which we know
at once, as the size and brightness of the object and the
intervening objects all seen, that we determine distance
by the eye. Later physiological and psychological re-
search seem to be showing that in the exercise of the
senses there are organic processes and mental processes
deeper down than those which appear on the surface.
But whatever intermediary steps there be, there must
be beyond and beneath them, and this to start with, a
knowledge of body occupying space. Yet in order to
uphold this doctrine certain distinctions must be drawn.

SECTION III.

DISTINCTIONS TO BE ATTENDED TO IN HOLDING THE DOCTRINE OF NATURAL REALISM : EXTRA-MENTAL AND EXTRA-OR-GANIC KNOWLEDGE ; SENSATION AND PERCEPTION ; ORIGINAL AND ACQUIRED PERCEPTIONS.

EXTRA-MENTAL AND EXTRA-ORGANIC. — All knowledge obtained through the senses is discerned as extra-
mental, that is, as out of and beyond the perceiving

mind. Our perception of the organs of the body, say
the tongue or the eye, is of something not in the self
cognizing it. But we come to know objects outside our
perceived organs and affecting them. It is thus that on
stretching out our hand or foot we find something, a stone
or board, resisting ; this knowledge may be called extra-
organic. All our cognitions through the senses are extra-
mental; those through some of the senses, such as the
sight and the muscular sense, are also extra-organic, that
is, they look at objects beyond our bodily frame.

SENSATION AND PERCEPTION. — It may be noticed
that in all our knowledge through the senses there is another element, and that is feeling of some kind. When we
know our hand we may know it in a pleasant or unpleasant
state. We may and we ought to distinguish between the
two. We call the one Perception and the other Sensation. These always go together. There is never a sen
sation without a perception, say, a perception of our
organism or of an object affecting it. On the other hand,
there is never a perception without a sensation of some
kind, strong or faint, pleasant, painful, or indifferent.
The sensation seems to be a mental affection or feeling
of an organic state.

These two, the perception and sensation, have by no
means the same intensity. It very often happens that
when the perception is strong the sensation is weak, and
vice versa, when the sensation is powerful the perception
may all but disappear. Thus in listening to an instructive speaker, our attention may be fixed on his words,
of which we wish to ascertain the meaning; whereas in
listening to music our soul may be exclusively occupied
with the rich sounds. There is a sense in which the
two are in the inverse order, the one of the other. If
the feeling is very strong the object may be very much

lost sight of.   On the other hand, we may be so absorbed with the contemplation as scarcely to notice the associated sensation.   The soldier eagerly engaged in the fight with the enemy in front of him does not for a time feel the wound with which he is pierced.   In gazing at a historical painting, we may be so interested in the incident as not to notice the coloring ; whereas, in looking at a flower painting we so enjoy the rich hues as never to notice the disposition of the flowers.   This fact is an illustration of a more general law of our nature, that when we fix our attention on one part of a concrete or complex phenomenon presented, the other parts become dim, and may in the end very much vanish from the view.

ORIGINAL AND ACQUIRED PERCEPTIONS. — We have seen that man must have original perceptions. Such are those of savors by the taste, of odors by the nostrils, of sounds by the ear, of a colored surface by the eye, and of resistance by the muscular sense.   Unless we get these by an original inlet we can never acquire them by any derivative process.   A man born blind cannot form any understanding or idea of color ; it is Locke who tells us of the blind man who, on being asked what idea he had of the color of scarlet, replied that he thought it to be like the sound of a trumpet.   But then by combining our experiences and by reasoning from them we can add indefinitely to our knowledge.   Thus it is believed that originally human beings cannot estimate distance by sight, and yet it is mainly by this sense that the mature man is able to tell the distance of objects from one another and from himself.   He has acquired a knowledge of nearness or remoteness by the muscular sense, say by the hand pressing along a surface ; but now he is able by the eye discerning the shade of color or the apparent size to determine the distance of an object. In this acquired knowledge there is first an accumulation of experiences, and then an argument founded on them. We shall show that it is by drawing the distinction between our original and acquired perceptions that we are able to account for the apparent deception of the senses. Our original perceptions never deceive us, but, in the haste of observation and the rapidity of reasoning, we may pronounce erroneous judgments on our acquired perceptions.

SECTION IV.

THE SENSES; GENERAL REMARKS.

I begin my exposition of these by one or two remarks bearing on them all.

The sensation and the perception of the sensation have their seat not in the organs of sense, so called, or in the nerves attached to them, but in the brain. The palate, the nostrils, the ear, the touch, the eye might all be affected in a regular manner, but there would be no taste, smell, hearing, feeling, nor seeing, unless the action went up into the cerebrum. Attempts have been made to give a separate place to each of the senses in the brain. I deem it proper, without committing myself, to give the views on this subject of Professor Ferrier of London, who has localized the senses. It would appear that rays of light might reach the eye, pass through the coats and humors on to the retina and the optic nerve, and yet no object be seen if the movement did not go on to the local centre of sight. Seeing is not in the eye but in the brain.

Each sense gives its own sensation and perception. If the optic nerve is struck, light may be emitted; if the auditory, a sound is heard. But one sense cannot be made to give the impression produced by another.

Great aid is imparted to all the senses by motion. This was dwelt upon by Aristotle, and has since been noticed by nearly all physiologists. Were the eyeballs motionless our knowledge of objects would be attained much more slowly and would be much more confined. We get a great increase of information by moving our sense organs, our eyes, our nostrils, or ears, so as catch different impressions. We would have a very vague idea both of space and time without locomotion.

SECTION V.

ORGANIC AFFECTIONS.

Those of the nerves of the internal organs of the body, such as the stomach, the alimentary canal, and the viscera, also of the physiological acts of respiration, digestion, breathing, and circulation, and specially of temperature, which though intimately allied to feeling is yet separate, may first be considered.

Each of these furnishes a peculiar sensation. The feelings from the whole are very numerous and very varied, and may constitute a considerable portion of human pleasure or human suffering. Such is the comfort produced by our bodily wants being supplied by air and water and food, and the stimulating cheerfulness arising from perfect bodily health. Such are the nervous affections, painful or pleasant, exciting or dull, irritating or soothing, depressing or elevating; and the uneasiness or pain coming from a diseased bodily frame. In all such affections the main element is sensation, but mingled with .t, though often very faint, is a perception of the part affected. This is not of any object, extra-organic. We may, however, by experience and reasoning come to know that this pain proceeds from a wound produced by a blow

or from an unhealthy atmosphere. But the original perception is only of an affection of our body of which we
know the direction and in a loose way the locality. But
upon these simple original perceptions we may rear a
body of acquired ones. We may come to know, for example, what kinds of food and air derange our systems
and what kinds stimulate or strengthen us. The affections of which I have been speaking constitute a sort of
general sense, which seems to be strong in some of the
lower animals.

The visceral affections are localized by Ferrier in the occipital
lobes of the brain (Fig. 1, D, D′, D″). When this part of the brain
is injured the animal will have no relish for its food and will not seek
for it. This sense becomes differentiated into special senses.

SECTION VI.

TASTE.

Its seat is in the upper surface of the tongue, which is
covered with papillæ of different kinds, and is supplied
with two nerves, the glosso-pharyngeal and the gustatory, a
branch of the fifth pair. The matter affecting the tongue
must be in a liquid state in order to its being felt. Taste
is affected by mechanical means, as by irritating the root
of the tongue. Many seeming tastes may be regarded as
smells ; e. g., an onion and an apple, if the nose be closed,
cannot be distinguished from each other by taste.

The sensations furnished are considerably diversified,
and cannot be classified very accurately or properly des-
ignated as they run into each other. Some are keen and
some insipid, some sweet and some bitter, some luscious
and some acrid. In this sense the sensation is far more
powerful than the perception. Still the perception is always present. We have a vague knowledge of the taste

being in our mouth and in a certain relative direction. Acting on the principle of causation we seek for a cause of the sensation, and by observation we may find it to be some kind of meat or drink, and by a gathered experience determine what kind of food it is. Some have ac-

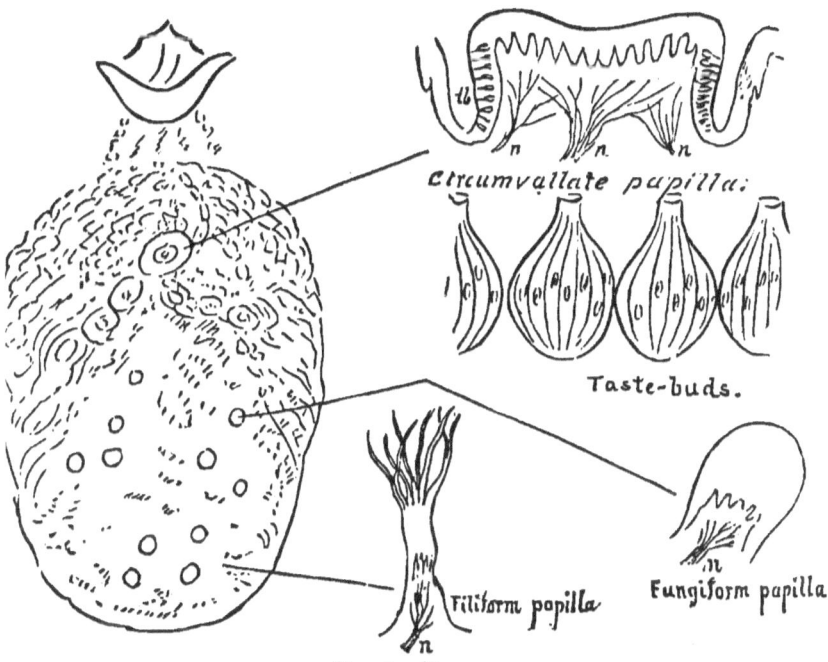

FIG. 3. TASTE.

DIAGRAM OF THE TONGUE, showing the *circumvallate papillæ*, enlarged, with their nerves (*n, n, n*), and taste-buds, also the *fungiform papillæ* and the *filiform papillæ*, and nerves (*n, n*) entering them. The nerves to the papillæ are branches of the glosso-pharyngeal nerve. It has been recently found that the so-called taste-buds occur on parts of the mouth which have no sense of taste.

quired a great delicacy in distinguishing the qualities of such articles as wine and tea. But there is no evidence that by this sense we know originally and intuitively anything beyond our frame. The knowledge is of objects extra-mental but not extra-organic.

3

SECTION VII.

SMELL.

Its organ is the nose, and the sensibility is in the mucous membrane lining the upper part of the interior and the cavities which branch from it. It has a special pair

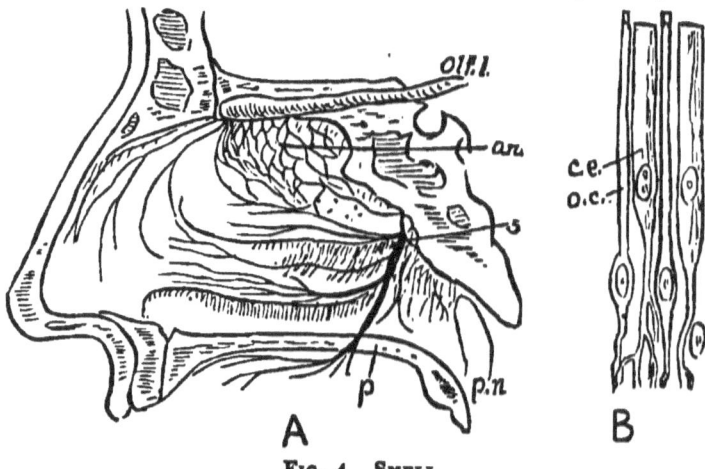

FIG. 4. SMELL.

DIAGRAM OF NOSE. A, showing olfactory lobe (*olf. l.*) from brain, with its olfactory nerves (*o. n.*); 5 is a branch of the fifth or trigeminal nerve; it sends branches to the lower region of the nose, and also to the palate; *p.*, palate; *p. n.*, posterior nares (where the nose opens into the mouth). B shows the fine olfactive cells (*o. c.*), ending in soft processes on the epithelium of the nostrils. They alternate with columnar epithelial cells (*c. e.*)

of nerves, or rather processes of the brain — the olfactory. An olfactory lobe of the brain proceeds to the region above each nostril and sends down olfactory nerves into the upper part of the nostril. These nerves supply rod-shaped epithelial cells, some wide, some narrow. The lower part of the nostrils is supplied by nerves of common sensation from a branch of the fifth or great trigeminal nerve.

The matter affecting the nostrils must always be in a gaseous state, and is called odor. Odors are so varied that we have not specific names for them; we speak of them as sweet, fresh, ethereal, stimulating; and of mal-odors as acrid, nauseous, disgusting. Smell is closely connected with taste. Both seem to be combined in fla-vor. Often, by combining the two, we have to determine the nature and state, whether sound or corrupt, of the food presented to us. Smell always contains perception, a perception of our nostrils as affected, but the sensations are always more predominant. All that we know imme-diately by this sense seems to be our affected organism. If the odor is one unknown, we have no idea of the ob-ject from which it comes. The senses of taste and smell are the most animal of the senses. Yet smell is capable of imparting a considerable amount of information, es-pecially of direction. Some of the lower animals seem to be guided in their movements by this sense. By it the dog will follow the track of game or of its master, or that which it has gone over itself previously, with won-derful accuracy. As we ascend the scale of animals, this sense seems to lose its importance and its acuteness. But by it our acquired perceptions carry us a consider-able distance beyond our bodily frame, and open to us a wider world than taste does. Smell and taste are sup-posed to have their centres not easily distinguishable in the Subiculum Cornu Ammonis. (Fig. 2, K., p. 6.)

SECTION VIII.

HEARING.

In this sense we have both sensation and perception in about the same proportion, though sometimes the sensa tion is the stronger, as in music, and sometimes the per ception, as when we are listening to the words of a speaker. It gives, primarily, a knowledge of our ears as affected, but by a combined experience we perceive objects at a distance, and know that this sound proceeds from the human voice, this other from a drum, or from the wind agitating the trees, or from a running stream. The organ of hearing is the ear collecting the sound,

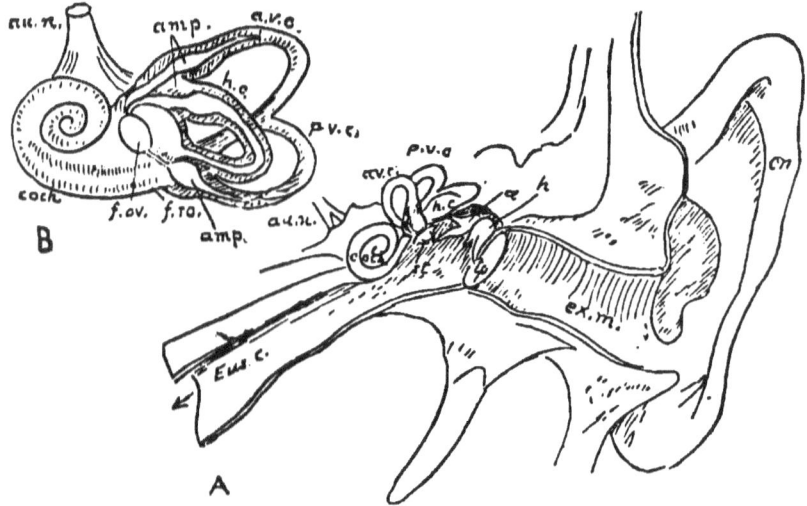

FIG. 5.　HEARING.

DIAGRAM OF LEFT EAR. A, general arrangement of parts. B, inner labyrinth (enlarged) *a.*, anvil; *amp.*, ampullæ; *au. n.*, auditory nerve; *a. v. c.*, anterior vertical canal; *cn.*, concha, or outer ear; *coch.*, cochlea; *Eus. c.*, Eustachian canal; *ex. m.*, external meatus; *f. ov.*, foramen ovale; *f. ro.*, foramen rotundum; *h.*, hammer; *h. c.*, horizontal canal; *p. v. c.*, posterior vertical canal; *st.*, stirrup.

the middle ear or tympanum with its bones or muscles, and the internal ear or labyrinth, presenting a spiral shell called the cochlea, and the semicircular canals, and containing a clear liquid. The matter affecting the organism is in a state of vibration. Going in by the external ear, the vibrations strike a membrane, the tympanum, and are transmitted to a chain of bones. The stirrup bone communicates beats to the opening of the labyrinth, and compresses the liquid, and this affects the auditory nerve, which carries on the action to the brain. Each bag of the labyrinth is filled with fluid, and floats in fluid. It contains mobile ear-stones, that beat like pebbles on the ciliated epithelium, which is richly supplied with nerves. The semicircular canals are engaged in maintaining our equilibrium. Through them rapid rotation of the body causes vertigo.

Auditory sensations are more delicate and agreeable than those furnished by any of the other senses, and differ

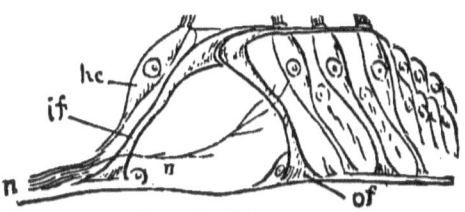

FIG. 6. HEARING.

DIAGRAM OF FIBRES OF CORTI : h c, hair cell ; i f, inner fibre ; n, nerves ; o f, outer fibre.

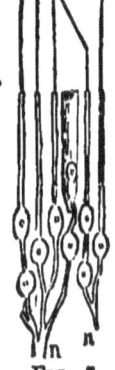

FIG. 7.

Hairs supplied with nerves (n) in ampullæ of ear. The vibrations of the fluid move the hairs.

in intensity, in quantity, and in tone. The melodies and harmonies of music stir up emotion, and by their continuance trains of emotional thought. The ear can appreciate very nice differences of sound, and the intellect is roused to interpret the articulate sounds of the human voice. The fibres of Corti are situated in the cochlea. They are said to have 6,000 inner, and 4,500 outer rods, and there are adjoining hair cells well supplied with nerves. They are usually supposed to be organs of music, and every tone affects a proper key of Corti's fibres. It seems certain that they somehow give the appreciation of sounds. They enable us to distinguish intensity of sounds and differences in time.

Hearing has its centre in the Superior Temporo-sphenoidal Convolution (Fig. 1, B). When this is destroyed there is no response
to the usual forms of auditory stimuli, such as calling, whistling, and
knocking.

According to a report by M. Elie de Cyon on the Semicircular
Canals and the Sense of Space (see "Mind," October, 1878): (1.)
Through the semicircular canals we obtain a series of unconscious
sensations bearing on the position of the head in space. (2.) Each
canal has a strictly determinate relation to one of the dimensions of
space. (3.) The loss of movement observed upon section of the canals
is due to the disturbance of the normal sensations of which they
are the organs." It is said we possess in the semicircular canals an
organ "fitted to form a notion of a space in three dimensions." The
semicircular canals are the peripheral organs of the sense of space;
that is to say, the sensations created through the nerve endings in
the ampullæ of the canals serve to form our notions of the three dimensions of space, the sensations of each canal corresponding with
one of the dimensions. By means of these sensations, there is
formed in our brain the representation of an ideal space, to which
are referred all the perceptions of objects around us, and the position of our body among these objects. The nature of the ideas
given by this apparatus needs to be carefully sifted.

SECTION IX.

TOUCH PROPER, OR FEELING.

In it we have sensation and perception more intimately
connected than in any other sense. The sensation arises
from the sensor nerves, proceeding from every part
of the periphery of the body to the sensorium in the
brain. The organ is the skin, and touch is often called
the skin-sense by the Germans. The skin consists of two
layers, the outer or cuticle, which is meant for protection
and is insensible, and the true skin, with its sensitive
points called papillæ lying under. Remove the epithelium and the sense of touch and that of temperature are
lost. The most sensitive parts of the body are the tips
of the tongue, of the fingers and the lips ; this is probably

because of the nerves generated at these points by use. The sensibility may be created from within, but is commonly awakened by pressure from without, which affects the papillæ and associated nerves.

We are led naturally (the nature may have been acquired by heredity) to refer the action to the point at which the sensor nerve terminates. If we prick a nerve which reaches the mid-finger, the pain is felt there. If we stretch or pinch the ulnar nerve by pushing it from side to side or compressing with the fingers, the shock is felt in the part to which its ultimate branchlets are distributed, namely, in the palm and back of the hand and in the fourth and fifth fingers. According as the pressure is varied the prick-

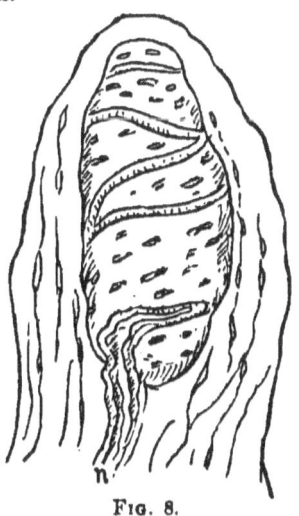

Fig. 8.

DIAGRAM OF TACTILE CORPUSCLE OF FINGER, with nerves (n) entering it.

ing sensation is felt by turns in the fourth finger, in the fifth, in the palm of the hand or back of the hand; and both on the palm and on the back of the hand the situation of the pricking sensation is different according as the pressure on the nerve is varied ; that is to say, according as different fibres or fasciculi of fibres are more pressed upon than others. The same will be found to be the case in irritating the nerve in the upper arm " (Müller's " Physiology " by Baley, p. 740). So strong is this tendency to localize the sensation at the extremity of the nerves, that when an arm or leg is amputated the person has still the feeling of the lost limb. Müller has collected a number of such cases. " A student named Schmitz had his arm amputated above the elbow thirteen years ago; he has never ceased to have sensations as if in the fingers. I applied pressure to the nerves in the stump, and M. Schmitz immediately felt the whole arm, even the fingers, as if asleep." "A toll-keeper in the neighborhood of Halle, whose right arm had been shattered by a cannon-ball in battle, above the elbow, twenty years ago, and afterwards amputated, has still, in 1833, at the time of changes of the weather, distinct rheumatic pains which seem to him to exist in the whole arm, and though re-moved long ago the lost part is at those times felt as if sensible to draughts of air. This man also com-pletely confirmed our statement that the sense of the integrity of the limb was never lost." When there is a change made artificially in the peripheral extremities of nerves, the sensations are felt as if in the original spots. "When in the restoration of a nose a flap of skin is turned down from the forehead and made to unite with the stump of the nose, the new nose thus formed has, as long as the isthmus of skin by which it maintains its original nerve-supply remains undivided, the same sensations as if it were still on the forehead; in other words, when the nose is touched the patient feels the impression in the forehead. This is a fact well known to surgeons, and was first observed by Lisfranc."

Whatever it may have been originally, all this is now natural, very probably handed down by heredity. "Pro-fessor Valentin ('Repert für Anat. und Phys.' 1836, p. 330) has observed that individuals who are the subjects of congenital imperfection or absence of the extremities have, nevertheless, the internal sensations of such limbs in their perfect state. A girl, aged nineteen years, in whom the metacarpal bones of the left hand were very short and all the bones of the phalanges absent, a row of imperfectly organized, wart-like projections representing the fingers, assured M. Valentin that she had constantly the internal sensation of a palm of the hand and five fingers on the left side as perfect as on the right. When a ligature was placed round the stump she had the sensation of formication in the hand and fingers, and pressure on the ulnar nerve gave rise to the ordinary feeling of the third, fourth, and fifth fingers being asleep, although these fingers did not exist. The examination of three other individuals gave the same results." (Ib. p. 747.)

By the simultaneous sensations we have a perception of a plurality of points, and we feel our bodily frame, as it were, round about us. Müller maintains that in this way we get a knowledge of the greater number of the parts of the body and in all the dimensions of space, and that when our body comes into collision with another body, if the shock be sufficiently strong, the sensation of our body to a certain depth is awakened, and there arises a sensation of the contusion in the whole dimensions of the cube. If this be true, then this sense gives us a knowledge of our body as extended in three dimensions.

The primitive knowledge given by this sense seems, as in the case of taste, smell, and hearing, to be intra-organic, though of course extra-mental. But by experience we come to know that there are extra-organic bodies affecting us and the cause of the sensations, and may thus come with the aid of the muscular sense to be cognizant of the hardness and softness, roughness and smoothness of the bodies touching us. It seems now to be ascertained that temperature is an affection of the tactile nerves. It is felt as a sensation of our bodily organs. The extra-organic cause is determined by experience and reasoning upon it.

Touch has its centre in the Hippocampal region. (Fig. 2, K, p. 6.) " Destructive lesions of this region abolish tactile sensation on the opposite side of the body."

SECTION X.

THE MUSCULAR SENSE.

This is intimately connected with Touch Proper, but differs from it essentially. The organic apparatus consists first of a motor nerve, proceeding from the lower part of the brain to a muscle. We will to move an organ, say the arm, and the motor nerve carries the action to the muscle. This part of the process has been called the Locomotive Energy. We know that the muscle has been moved and resistance offered, by a sensor nerve attached and carrying the intimation to the brain. In this sense as in every other there is sensation, but perception is vastly predominant. By the senses which have come before us hitherto, we seem merely to know our frame with its parts out of each other. By this sense we know objects out of and beyond our body and as resisting our energy. The senses already noticed have given us linear direction, probably also surface, plane or perhaps curved ; they have certainly given us points of space as separated ; this gives us bodies in three dimensions. We press on a solid body and along its surface, and along its sides and around it, and thus get the idea of solidity or impenetrability. The muscular sense, including in it the volition and the resistance, first gives us the idea of Power, Potency, Energy, or Force, out of which proceeds our idea and conviction as to Causation.

While Feeling and the Muscular Sense are different, the one being intra-organic and the other extra-organic yet they commonly act at the same time and together. They unite to give us the sense of *pressure* which arises from the force with which a body presses on our nerves of feeling and is resisted by muscular action. A body is laid on our skin and we estimate its *weight* by the amount of force which we use in order to lift it. By practice people may become very expert in weighing objects. Those who have to mix materials in definite proportions can often do so without the use of a machine, and the officers in a post-office can tell the weight of a letter by simply placing it on their hand.

SECTION XI.

VISION.

This is in many respects the highest and most intellectual of all the senses. It is also the most complicated. I am not sure that all its mysteries have yet been cleared up. Much, however, is known. We have to contemplate it simply as giving us a perception.

The ball of the eye is a globe moving freely in a chamber, the orbit. It has a firm, tough, spheroidal case, the greater part of which is white and opaque, called the sclerotic. In front is what is called the cornea, which is transparent. Light enters by the cornea, and thence passes into the aqueous humor, consisting mainly of water. It then passes through the gateway of the iris into the denser crystalline lens, where it is refracted according to the shape and consistency of the lens. It now passes through the vitreous humor, which is a sort of jelly, to the retina, where it forms an inverted image of the object from which it has come. On the retina it impacts on rods and cones which are connected with the optic nerve. The estimated number of cones in the human eye is 3,360,000; the number of rods is not known. The rods have a pigment which is bleached by light and restored in darkness. We do not know the full functions of these rods and cones; they seem, however, to be connected with the formation of the figure,

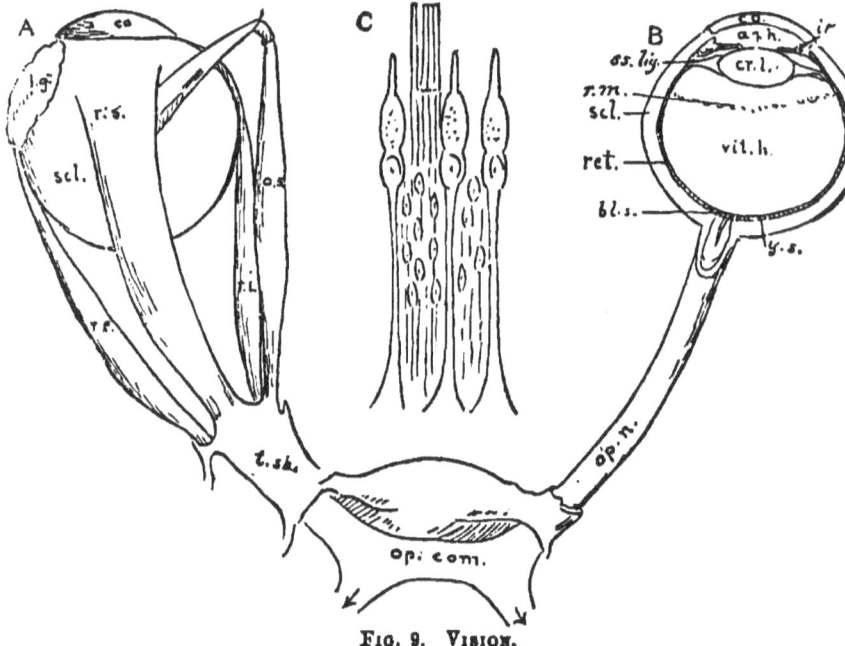

FIG. 9. VISION.

DIAGRAM OF EYES. A, left eye-ball, showing the muscles; B, right eye in section; C, section of retina, magnified, showing rods and cones. *aq. h.*, aqueous humor; *bl. s.*, blind spot; *co.*, cornea; *cr. l.*, crystalline lens; *ir.*, iris; *l. g.*, lachrymal gland; *op com.*, optic commissure (the arrows mark the course of the optic tracts to the brain); *op. n.*, optic nerve; *o. s.*, superior oblique muscle; *r. e.*, rectus externus muscle; *r. i.*, rectus interuus; *r. m.*, retinal margin; *r. s.*, rectus superior muscle; *ret.*, retina; *scl.*, sclerotic; *ss. lig.*, suspensory ligament of lens; *t. sh.*, tendinous sheath of nerve; *vit. h.*, vitreous humor; *y. s.*, yellow spot (where vision is most distinct) The points of the rods and cones at C are directed *backward* in the retina.

certainly of the color. There is no vision at the point where the light falls on the optic nerve, and it is called the blind spot, which has no cones or rods. Vision is most acute at a yellow spot which is full of close-set

cones. Distinctness of vision requires that objects shall
be so far apart that their images on the retina shall reach
more than one cone. The luminous action remains not
only during the time the light is shining, but an appreciable time after. The retina in some persons seems to
be affected in the same way by various colors. This
gives rise to color-blindness, so that the person cannot
distinguish between the green leaves of a tree and its
red fruit.

There are large muscles, straight and oblique, which
keep the eye in its place and direct its axis, so that we
can carefully gaze on and inspect the object. Were the
eye-ball fixed, our knowledge by the eye would be very
imperfect. Motion in this sense greatly helps us in our
perception of objects.

Intuitively we perceive by the eye a colored surface,
and I believe nothing more. This surface is felt as affecting us. But by a gathered experience and reasoning
upon it, we can extend our knowledge indefinitely. It
had been surmised by several persons before, as by Locke,
but was established by Bishop Berkeley, the Irish metaphysician, in " New Theory of Vision " (1709), that originally we have no knowledge of linear distance by the
eyes. On looking forward we have simply a perception
of a colored surface affecting us, at what distance we cannot tell. This theory has since been confirmed by the observation of the cases of persons born blind, but whose ·
eyes were subsequently couched so that they could see.
I shall mention three of these cases.

Cheselden Case. — The boy was between thirteen and
fourteen years of age when his eye was couched by Dr.
Cheselden (see " Trans. of Royal Society," 1727). When
he first saw, he was so far from making any judgment
about distances, that he thought all objects whatever

touched his eyes (as he expressed it), as what he felt did
his skin, and thought no objects so agreeable as those
which were smooth and regular, though he could form no
judgment of their shape or guess what it was in any object that was pleasing to him.   He knew not the shape
of anything, nor any one thing from another, however
different in shape or magnitude, but being told what
things were, whose form he before knew from feeling, he
would carefully observe that he might know them again ;
but having too many objects to learn at once, he forgot
many of them, and (as he said) at first learned to know
and again forget a thousand things in a day.   One particular only, though it may appear trifling, I will relate.
Having often forgot which was the cat and which the
dog, he was ashamed to ask, but catching the cat, which
· he knew by feeling, he was observed to look at her stead-
fastly, and then putting her down said, " Puss, so I shall
know you another time."   We thought he soon knew
what pictures represented which were shown him ; but
we found afterwards we were mistaken, for about two
months after he was couched he discovered at once they
represented solid bodies, when to that time he considered
them only as party-colored planes or surfaces, diversified
with variety of paints ; but even then he was no less surprised, expecting the pictures would feel like the things
they represented, and was amazed when he found those
parts which by their light and shadow appeared now
round, and even felt flat like the rest ; and asked which
was the lying sense, feeling or seeing."

Franz Case (" Phil. : Trans. of Royal Society," 1841).
The youth had been born blind and was seventeen years
of age when his eye was couched by Dr. Franz, of Leipsic.   When the eye was sufficiently restored to bear the
light, " a sheet of paper on which two strong black lines

nad been drawn, the one horizontal, the other vertical, was placed before him at the distance of about three feet. He was now allowed to open the eye, and after attentive examination he called the lines by their right denominations." " The outline in black of a square, six inches in diameter, within which a circle had been drawn, and within the latter a triangle, was, after careful examination, recognized and correctly described by him." " At the distance of three feet, and on a level with the eye, a solid *cube* and a *sphere*, each of four inches diameter, were placed before him." " After attentively examining these bodies, he said he saw a *quadrangular* and a *circular* figure, and after some consideration he pronounced the one a *square* and the other a *disc*. His eye being then closed, the cube was taken away and a disc of equal size substituted and placed next to the sphere. On again opening his eye he, observed no difference in these objects, but regarded them both as discs. The solid cube was now placed in a somewhat oblique position before the eye, and close beside it a figure cut out of pasteboard, representing a plane outline prospect of the cube when in this position. Both objects he took to be something like flat quadrates. A pyramid placed before him with one of its sides towards his eye he saw as a plain triangle. This object was now turned a little so as to present two of its sides to view, but rather more of one side than of the other: after considering and examining it for a long time, he said that this was a very extraordinary figure; it was neither a triangle, nor a quadrangle, nor a circle; he had no idea of it, and could not describe it; ' In fact,' said he, ' I must give it up.' On the conclusion of these experiments, I asked him to describe the sensations the objects had produced; whereupon he said that immediately on opening his eye he had discovered a difference

in the two objects, the cube and the sphere, placed before him, and perceived that they were not drawings;
but that he had not been able to form from them the
idea of a square and a disc until he perceived a sensation
of what he saw in the points of his fingers, as if he really
touched the objects.   When I gave the three bodies (the
sphere, cube, and pyramid) into his hand, he was much
surprised that he had not recognized them as such by
sight, as he was well acquainted with mathematical figures by his touch."   " When the patient first acquired
the faculty of sight, all objects appeared to him so near
that he was sometimes afraid of coming in contact with
them, though they were in reality at a great distance from
him.   All objects appeared to him perfectly flat ; thus,
although he very well knew by his touch that his nose
was prominent and the eyes sunk deeper in the head, he
saw the human face only as a plane."

These observations show that the eye takes in surface
and superficial figure at once, but cannot discern solidity.
If the persons have the use of both eyes, they will observe the difference between a disc and a solid, but they
would not be able to say till they feel it that the latter
is a solid.   It requires to be added, that those who have
their sight thus given them require observation and
thought to reconcile the information they had got from
touch with that which they are now receiving from sight ;
just as people who have learned two languages, say German and French, require practice in order to enable them
readily to translate the one into the other.

Another portion of this report is worthy of being recorded, as showing how the memory and the fancy depend
on the senses.   " Though he possessed an excellent memory, this faculty was at first quite deficient as regarded
visible objects ; he was not able, for example, to recog-

nize visitors, unless he heard them speak, till he had seen them very frequently. Even when he had seen the object repeatedly, he could form no idea of visible qualities in his imagination without having the real object before him. Heretofore, when he dreamed of any persons, of his parents, for instance, he felt them and heard their voices, but never saw them ; but now, after having seen them frequently, he saw them also in his dreams."

Trinchinetti Case. — Mr. Abbot (in " Sight and Touch ") gives an account of the observations of Trinchinetti : " He operated at the same time on two patients (brother and sister), eleven and ten years old respectively. The same day, having caused the boy to examine an orange, he placed it about one metre from him and bade him try to take it. The boy brought his hand close to his eye, and closing his fist found it empty, to his great surprise. He then tried again a few inches from his eye, and at last. in this tentative way, succeeded in taking the orange, When the same experiment was tried with the girl, she also at first attempted to grasp the orange with her hand very near the eye, then, perceiving her error, stretched out her forefinger and pushed it in a straight line slowly until she reached her object." Trinchinetti " regards these observations as indicating that visible objects were in actual contact with the eye." Other patients have been observed (by Janin and Duval) to move their hands in search of objects in straight lines from the eye.

But while the perception of distance is not an original endowment of sight, it can be acquired. It should be noticed that in this acquisition we are much aided by the circumstance that while we do not by the eye perceive distance from us, we see a flat surface with a distance between the sides.

Means by which we are able to estimate Distance by the

4

Sense of Sight. — For near objects there are three special aids provided in the organism itself, and there are others for more distant objects.

(1.) When we look at near objects the pupil slightly contracts, and the anterior surface of the crystalline lens becomes more convex. The process by which this is done is a somewhat complex one, in which there is probably both reflex and voluntary action. As it takes place there is a strain in the action of the eye, and intimation is given of this by the attached nerves. When this strain is felt we know by experience that the object is near.

(2.) There is a difference of the parallelism of the rays of light according as the objects are near or remote. When objects are at a distance the rays that come from them are virtually parallel, and the eye keeps its normal shape in receiving them. But when objects are near the rays are not parallel even approximately, and the eyes are strained in taking them in. Announcement of this is given to the mind, not by the eye-balls directly, but by the attached muscles. We come to argue that the object is near when the muscles are strained.

(3.) There is a difference, according as the object is remote or near, of the image produced on the retina by each of the two eyes. When the object is at a distance the figure given by the two eyes does not differ much from that produced by one. But when it is near there is a sensible difference. Place the back of a closed book before the eyes, twenty feet away, and there will be little difference between the form as given by two eyes and by one. Place it a foot away, and we see much more of the two sides by the two eyes than by one. There are other means which apply to objects at all distances.

(4.) There is the difference of relative size of the felt impression on the retina, as the objects are near or distant. A penny placed close to the eye may occupy the whole field of vision, may, according to the proverb, hide the sun from the view. Place it at some distance and it will occupy a comparatively small space in the figure painted on the retina.

(5.) When an object, say a watch, is at a distance, the rays of light that come from it produce a much feebler impression on our organism than when it is near. We argue that an object is far off when its color is faint and its outline hazy. We infer that it is near when its color is bright and its figure distinct.

(6.) In our reasonings about the distance of objects we are much guided by the number of intermediate objects on which the eye can rest. When these objects are numerous we conclude that the object must be at some distance, and when they are few we are apt to argue that it must be near. This rule often enables us to guess very rapidly at the distance of objects. On the other hand, as we shall immediately see, it may often lead us into error by being illegitimately applied.

(7.) We are often guided in our estimate of the distance of an object by its known size. The object, let me suppose, is evidently a human being, a man or woman, and occupies a certain place in the retinal affection. The image is very small and we conclude that the object, man or woman, must be at a distance. Or, it is large, and we infer that the object is close to us.[1]

When both eyes are in healthy exercise there is a double image on the retina. But the object is seen

[1] "I shall say nothing," says Sidney Smith, "of the moral method of measuring distances; the distance from home to school in the days of our youth being generally double the distance from school to home, and so with all other passages which quicken or retard the feeling of time."

though there be only one image. The object is perceived as single when the images are thrown on the proper parts of the retina. When they are not so, the object may be double or misplaced.

The image on the retina is inverted. The arrow with the point up has the point down in the retinal image; yet the object is seen upright. This circumstance has puzzled many. The puzzle arises from the circumstance that people imagine that there must be an inner eye of some kind looking at the retinal image; whereas that image is not seen by any but the physiologist pursuing his researches. It is, in fact, a mere mechanism, or means to let us know the shape and direction of the object; and it is governed by the law of visible direction, which is, when the rays strike the retina we trace them back along the line by which they have come. The rays at the base of the retinal figure have come from the top of the object, say an arrow, and we place them at the top, while those at the top have come from the foot, thus giving the object its real position.

We are now in the heart of a subject which deserves a brief separate consideration.

SECTION XIL

OUR ACQUIRED PERCEPTIONS.

They are acquired by a gathered observation and by reasoning from this. In Taste our original perception is of the palate as affected, but we infer from repeated eases that this taste is caused by water and this by bread or by beef, and the perception by practice may become very acute. In Smell, we know at first only an affection of the nostrils, but we come to know by reasoning upon experience that this odor proceeds from a rose and this other from a lily, at this side or that side of us, according as it affects more strongly the right or the left nostril, and that the known smell must come from a near or remote object according to its intensity. In Feeling we seem to perceive intuitively only the periphery of our bodies, but we conclude that this agreeable sensation comes from a wholesome atmosphere, and this painful one from a blow or from excessive heat or cold. In Hearing we know directly our ear as affected, but we gather that the sound comes from the right when it is stronger in the right ear and from the left when it is more intense in the left ear; and that this sound is issued by a human voice, and this other by the wind or by a drum. By the muscular sense we may come to know very accurately the pressure implied in a blow, or the weight of an object lying on our hand or any other part of the body. Attention has been already called to the way in which we are able to estimate distance by sight. There are other acquired ocular perceptions which should be noticed.

We judge of the size of objects by comparison of them with other objects whose size we know. I see a plant unknown to me alongside a figure which I know to be that of a cow, and I determine the height of the plant because I am acquainted with the height of the cow. Proceeding on this principle, a painter, when he wishes us to appreciate the height of a building, or of a precipice, places a man or woman in front of it. If he wishes us to know that this animal is a foal, he places beside it a full-grown horse.

We can come to know the solidity of objects by means of binocular vision. Primarily, we become acquainted with the three dimensions of bodies by means of the muscular sense, by which we feel round them and grasp them. The eye, we have seen, perceives intuitively only

a colored surface, and a solid is noticed as a plane surface. No doubt it might see a sphere and a cube to be
different, but it would not discern the cube to be a cube.
But when a solid object is not remote, each eye gives a
different aspect of it. By combining the two perspec-
tives, we come to know the object as having three dimensions. Those who have but one eye make up for their
want by moving the head from side to side, so as to obtain
the same views as are to be had by the two eyes.

"Mr. Saunderson, the blind mathematician, could distinguish by his
hand, in a series of Roman medals, the true from the counterfeit,
with a more unerring discrimination than the eye of a professed vir-
tuoso, and when he was present at the astronomical observations in
the garden of the college, he was accustomed to perceive every cloud
which passed over the sun. This remarkable power, which has sometimes been referred to an increased intensity of particular senses, in
many cases evidently resolves itself into an increased habit of attention to the indications of all those senses which the individual retains.
Two instances have been related to me of blind men who were much
esteemed as judges of horses. One of these, in giving his opinion
of a horse, declared him to be blind, though this had escaped the observation of several persons who had the use of their eyes, and who
were with some difficulty convinced of it. Being asked to give an
account of the principle on which he had decided, he said it was by
the sound of the horse's step in walking, which implied a peculiar
and unusual caution in his manner of putting down his feet. The
other individual, in similar circumstances, pronounced a horse to be
blind of one eye, though this also had escaped the observation of
those concerned. When he was asked to explain the fact on which
he formed his judgment, he said he felt the one eye to be colder than
the other. It is related of Dr. Moyse, the well-known blind philos-
pher, that he could distinguish a black dress on his friends by its
smell, and there seems to be good evidence that blind persons have
acquired the power of distinguishing colors by the touch. In a case of
this kind mentioned by Mr. Boyle, the individual stated that black imparted to his sense of touch the greatest asperity, and blue the least.
Dr. Rush relates of two blind men, brothers, of the city of Philadel-
phia, that they knew when they approached a post in walking across a

street, by a peculiar sound which the ground under their feet emitted in the neighborhood of the post; and that they could tell the names of a number of tame pigeons with which they amused themselves in a little garden, by only hearing them fly over their heads. I have known several instances of persons affected with that extreme degree of deafness which occurs in the deaf and dumb, who had a peculiar susceptibility to particular kinds of sounds, depending apparently upon an impression communicated to their organs of touch or simple sensation. They could tell, for instance, the approach of a carriage in the street without seeing it, before it was taken notice of by persons who had the use of all their senses. An analogous fact is observed in the habit acquired by the deaf and dumb of understanding what is said to them by watching the motion of the lips of the speaker." (Abercrombie's " Intellectual Powers.") " An American Indian has such acute sight that he can discover the prints of his enemies' feet, can ascertain their number with the greatest exactness, and the length of time which has elapsed since their passage; he can discover the fires and hear the noises of his enemies when no sign of the contiguity of any human being can be discovered by the most vigilant European." (Smith's " Moral Philosophy.")

SECTION XIII.

APPARENT DECEPTION OF THE SENSES.

The Greek philosophers, down to the time of Aristotle (who corrected the mistake), represented the senses as deceiving us. The distinctions we have drawn, especially that between our original and acquired perceptions, enable us to stand up for the trustworthiness of our sense-perceptions. Our original perceptions are all true to facts; but there may be mistakes in the steps we take in forming our derivative perceptions. Our observations may be limited, and we may argue from them as if they were unlimited. The taste in the mouth, as a mere organic affection, is always what we may feel it to be; but we may draw a wrong inference as to the object in the mouth, as to whether it is beef or mutton, as to whether it is sherry or madeira wine; and when our palate or stomach is deranged, we may regard sound meat as unsound. We cannot be mistaken in regard to the smell as a sensation, but we may err in our conclusion that it is produced by a certain object in a certain direction at a certain distance. For our convenience we lay down rules for our guidance as to the objects falling under the senses, which are correct enough for ordinary purposes, but fail and mislead us in exceptional circumstances. Sounds come to our ears in straight lines, but the sound coming from a bell may be diverted by a building in the way, and we trace the sound to the direction from which it has last come. A man with an amputated limb places the pain in it, because it is precisely what he would have felt if the limb had been entire.

The supposed illusions are most numerous in the use of the sense of sight, and this because there are so many observations and ratiocinations implied in our judgments in regard to the position and distance of objects by that sense. We are accustomed to estimate distances of an object by the number of visible objects coming between us and it; and we are apt when we are looking across a lake or an arm of the sea, a level plain or a waste of sand, to regard them as much nearer than they are. We are apt to draw a wrong inference when things are seen across a surface of snow. " We had frequent occasion," says Captain Parry, "in our walks on shore to remark the deception which takes place in estimating the distance and magnitude of objects when viewed over an unvaried surface of snow. It was not uncommon for us to direct our steps towards what we took to be a large mass of stone at the distance of half a mile from us, but which we were able to take up in our hands after one minute's walk. This was more particularly the case when ascending the brow of a hill." In all these reasonings we start from an assumed position and may proceed illegitimately. When he feels himself to be at rest on the deck of a ship which may in the meanwhile be starting from the shore, the countryman starts up in alarm, for he believes, momentarily, that the shore is moving. When we are looking out of a railway carriage on a train starting, we may feel as if we are moving, because the carriage we are looking at seems stationary, and we are not assured of the contrary till we see it passing an object which we know to be stationary, when, be it observed, we at once accommodate ourselves to the actual position. "I remember," says Abercrombie, "having occasion to pass along Ludgate Hill, when the great door of St. Paul's was open and several persons were standing in it. They appeared to be very little children, but on coming up to them they were found full-grown persons. In the mental process the door had been taken as of a certain magnitude (much less than it actually was) and the other objects were judged by it." In a mist the boy seems a man and the man a giant, because in our common experience the objects seen so dimly are at a distance, and this boy or man being at such a distance must be very large to fill such a space in our eye.

SECTION XIV.

SUPPLEMENTARY NOTES.

NOTE I.

AS TO WHAT WE PRIMARILY PERCEIVE.

Physiologists are seeking to find out the organic processes involved in the exercises of the Senses. Psychology should seek to determine what is the primary exercise of the conscious mind in Sense-Perception.

Certain German savans have been making diligent inquiry into the nature of the organic processes. Weber made some curious experiments as to the relative sensibility of different parts of the body, showing how much more sensitive the tip of the tongue is than the back. Lotze has been experimenting and speculating as to the origin of our notion of space, and discovers in each of the senses *local signs* which indicate the difference of an impression from others. According to my view all these local signs are in the organism, and are acknowledged to be movements there, and are at the best the mere prompters of the notion of space, and do not contain in themselves the notion of space or any other idea whatsoever. Fechner, in his "Psychophysic," has sought to determine the relation of the exciting cause to the sensation, and thinks he has proven that the sensation is not directly as the excitation, but the sensation increases as the logarithm of the excitation. Delboeuf and Hering dispute the conformity of this law to facts. It is certain, I think, that the law is a physiological and not a psychological one, is a law of the organism and not of the conscious mind. Wundt regards external impressions as mere signs to be interpreted; and maintains that they are interpreted by unconscious reasoning, which is the primary element of all thought. This view places reasoning prior to the notion and the judgment, which is contrary to the almost universal opinions of philosophers, and is supported by no evidence except that of a hypothesis of unconscious mental operations of which we have no proof. Helmholtz, who is a physicist rather than a metaphysician, divides the theories as to the origin of our ideas of space into nativist and empiricist. He opposes the nativist theory in the shape it takes in the philosophy of Kant, according to whom space is an *à priori* form in the mind imposed on objects. I do not believe in any such forms. According to the view expounded in this chapter the conscious mind has a native capacity of perceiving matter as presented to it. All these German theories may be modified if not set aside, if it be true, as Ferrier maintains, that each sense has an organ in the cerebrum, and that there is no perception unless the organic affection reaches the brain. Ferrier tells us that "on destruction of the *angular gyrus* the loss of vision is complete and permanent." (For the German theories see "La Psychologie Allemande Contemporaine," par Th. Ribot, translated by J. M. Baldwin.)

The microscope has not yet been invented which is fitted to show us the working of perception or any intelligent act of the mind. In order to get information we have now to employ, not the senses, but the consciousness. And there is a difficulty in determining what is the first conscious act. We cannot look into the soul of the infant when it is in the womb, nor for a considerable time after. It cannot express any of its affections except pleasure or pain, — say by a smile or a cry, — and we do not remember our early experience. In mature life it is found that the various physical and psychical acts are so mixed that it is difficult to separate them. Still, we can by self-consciousness look at our mental acts and observe what they are. We notice that in all of them there is a perception of an extended object within the organism or beyond it. Consciousness further testifies that in mature life we know matter as resisting our energy, certainly by the muscular sense, probably by all the senses. But neither of these can be had by reasoning or by development from a premise which does not contain them. They must therefore be given and not derived, intuitive and not acquired, premises and not a conclusion.

Let physiology penetrate as far as it can into the secrets of the organism, say in sight, into the structure of the eye, of the optic nerve, and it may be of the angular gyrus in the brain. But let it modestly stop when it comes to something which cannot be seen or touched, which cannot be weighed or measured. At that point let psychology take up the investigation and inquire what is the nature of perception, memory, reasoning, and other conscious acts. Physiology seems to declare that all that passes through the organism, through the nerves and brain, are vibrations. If it be asked, as has already been asked by Lotze, How can vibrations produce perceptions? I answer that the question of *how* (the διότι of Aristotle) is often difficult to answer. That there are vibrations is certain, that there are conscious perceptions is also certain. To determine their precise relation is acknowledged by all to be a perplexing question. The answer to it is not made easier by bringing in a *tertium quid* of any kind. If this medium is of the nature of matter, the question follows, How it can influence mind? If it is of the nature of the mind, How can it act on matter? If it is of the nature of neither, the unanswerable question is put, How can it operate both on mind and matter? While we cannot answer such questions, we can say that the conscious mind perceives matter as extended and solid. We may regard this as a native capacity of the cognitive mind until it is resolved into something simpler.

The most satisfactory position is that the mind perceives matter, that by all the senses it perceives the organism, and that by two of

the senses, sight and the muscular sense, it perceives objects affecting the organism.   Let us assume that perception is one of the capacities of mind, and probably we are as near the truth as we can
possibly be.   In the mature mind perception is a property of mind,
just as certainly as gravity is a property of matter or assimilation of
life.   As it cannot be derived from anything else, from material
action or vital action, we must regard it as original and primary.
We may assume that in it we perceive things as they are.   We perceive objects within or beyond our frame as extended and as affected.
True, we do not perceive the vibrations, which we know only by the
aid of science, but we perceive the affections produced by the vibrations.   These affections are in space, and the mind perceives them
as in space.   Thus a muscular action, say the movement of the arm,
is in space.   The affections of the palate, the nostrils, the ear, are
all perceived as in a certain direction and extended.   They are perceived as affections, as affected, as resisting.   We thus get at the
first perception, and in all subsequent perceptions of body, extension
and resisting power, which we may regard as the primary and universal properties of bodies.

NOTE II. ·

THE FOUNDATION LAID IN PHYSICAL NATURE FOR CONTINUED ACTION, FOR DEVELOPMENT AND YET FOR PERMANENCE.

Every bodily substance contains a certain capacity of energy; this
is quite as certain as that it contains a certain amount of particles.
This constitutes the basis of the conservation of energy, a doctrine
which follows from the nature of body when properly apprehended.
This energy is shown in one body acting on another by its properties.   The force operates when the conditions implied in its nature
are supplied.   A stone must fall to the ground if unsupported.   Hence
the perpetual changes in nature so fondly dwelt on by Heraclitus and
the Φιλόσοφοι Ρέοντες.

The forces in the agents which act as the causes are not lost.   In
all physical causation there are two or more agents in the cause.
In the action there is a change in each of the agents; for example,
both in the oxygen and hydrogen which combine to form water; but
the substances, the oxygen and hydrogen, abide with their capacities.
This is the τὸ ὄν of the Eleatics, which never changes.   There is thus
on the one hand, a "persistence of force," as Herbert Spencer calls it,
and at the same time a succession of actions.   This continuance with

mutation is evidently under a Divine order which takes the form of law. There is a sense in which all action is development, or evolution : the force comes out of the original energy in bodies. By their mutually adapted action, the forces often run in lines or races which are so arranged as to be periodical, they return according to their circuits, as for example, the seasons do, spring, summer, autumn, and winter, and the plant is after its kind.

SECTION XV.
ON THE EDUCATION OF THE SENSES.

The senses are all capable of being educated. Our tastes may be made more delicate, and may keep us from using deleterious food. The sense of smell may be cultivated, and add to our enjoyments ; and odors, especially by means of flowers, may be provided to gratify it. Hearing may be improved and made more sensitive and accurate. Music is a source of pleasure, which may be enhanced till it becomes elysian. Feeling may be made very delicate in its perceptions, and capable of distinguishing very nice differences of object. The senses of pressure and of weight may be so trained as to give us very accurate measurements. But the eye is the most intellectual of all our sense-organs, enabling us at a glance to take in the vast and the minute, the near and the distant.

All these should be cultivated by training in the family and at school. Children should be taught from their earliest years to use their senses intelligently and habitually. They should be encouraged to observe carefully the objects around them, and taught to describe and report them correctly. It has been said that there are more false facts than false theories, and this arises from persons not being trained to notice facts accurately, neither adding to them nor taking from them, nor gilding them by the fancy, nor detracting from them to serve an

and.   Pictures and models are used very extensively in
modern education, and serve a good purpose, as they call
in the senses to minister to the intellect.   But the things
themselves are vastly more instructive than any representations can be.   So children should be taught to use
their senses, especially their ears and their eyes, in observing the objects around them, and the events that
occur, and storing them up for future reflection.   Plants
and animals and stars, men and women and children, fall
under our eyes at all times, and their nature, shapes, and
actings should be diligently scanned for practical use and
for scientific attainment.   Not, indeed, that every fact
can be noted ; for this would lay a burden on the mind
which it cannot bear.   Pains should be taken not to dis-
tract the mind by too great an accumulation of details,
so as to prevent the rise and action of the reflective faculties.   But the habit of careful observation should be
acquired in early life, and facts stored up in all departments which we mean to study or to use in our future
lives.

SECTION XVI.

KNOWLEDGE GIVEN BY THE SENSES.

Having looked at the senses individually, let us now
weigh the results they yield when they are combined in
their action.

The Knowledge of Our Bodily Frame. — This acts, I
believe, as the starting-point of all our knowledge of
extra-organic objects, and furnishes a standard and a
measure.   Let us try to ascertain, in a general way, what
this combined organic knowledge amounts to.   By each
of the senses we have a knowledge of parts of our frame
as affected.   Already, then, we have a knowledge concrete of things as in space.   The part affected odorously

is in one direction; the part affected by hearing in another direction; that affected by color in a third; and so with the other senses; each sense localizes an organ, palate, nostrils, ear, eye, while touch proper gives a knowledge of the direction and locality of affections in every part of the body, and the muscular sense makes known the spot at which the energy is exerted. By combining this knowledge, we come to have a considerable and a familiar acquaintance with our bodily frame, with the parts, and affections thereof. It is, after all, however, very loose and imperfect, till we are able to perceive the body, as it were, *ab extra*, till we touch and handle it, and see the outside form of it. We know the shape of our bodies all the more distinctly from observing the figures of men and women around us. The peasant girl gains a large amount of interesting information when she sees her face and figure reflected in the water, or more perfectly in the mirror. When affected with the toothache, we know the general direction of the pain, but may not be able to tell in what tooth it is, as the same is known by the tongue or hand; it is certain that we cannot in this way know the form of the tooth. But, when we have toothache, we try to find out, by touch or sight, the tooth in which the pain is. This may illustrate the way in which we combine the intimations given by the different senses. As the result of all the steps, intuitive, experiential, and inferential, we carry with us always, and wherever we go, a sense of our circumambient body and of its several parts, of its capable acts and susceptible affections, and round this, as a nucleus, we gather information, and all our knowledge of objects beyond our frame is referred to this as the centre of our world.

Our Combined Extra-organic Knowledge. — At the very same time that we know our bodily frame we have a certain amount of knowledge of objects beyond it; we seem to take in a colored surface by the eye, and by the muscular sense we know objects as resisting our energy. Upon this foundation laid by nature we may rear an immense superstructure. Acquainted with the structure of the sensory organs, the boy is able to fix the direction of objects affecting them, of objects seen and touched; of that face which he sees, of that voice which he hears, of that arm that holds him, and he is soon able to trace them all to one person, his nurse or his mother. Thus do we fix the qualities discerned by different senses in one object. We smell an apple, we see its color and outline, we take it into our hands and feel its shape, we press it and ascertain its hardness, and we hear the sound the crushing makes. Henceforth the very smell or sight brings these qualities, or a number of them, before us, is associated with these qualities, and is conceived by us as possessing them. We expect everything that smells so, even when we do not see or touch it, to have a certain shape and consistency, and a certain taste in the mouth. We thus come to be surrounded by objects, with qualities attached to them, in our apprehension. We distribute objects in the room, doors, tables, chairs, desks, books, pictures. We know the place, and, so far, the properties of every object under our view in nature, of the trees, the fields, the meadows, the rivers, the clouds, the sun, moon, and stars. We learn by degrees the purposes served by the things before us. That object is a chair, with a piece of dress lying on it; that other is a table, with food on it; that other a horse, on which we may ride. As our observation and experience widen, our world enlarges; the known things in it become more numerous, and we know them more fully and accurately. In particular, we become acquainted with innumerable beings with like thoughts and sentiments as ourselves. We have at last not just a universe, but a cosmos with earth and air, plant and animal, with sun, moon, and stars, some of them at incalculable distances, and with innumerable living beings possessed of immortality. It should be noticed that all this knowledge radiates from our sensitive and conscious self. We place all these objects around us, in a certain direction from ourselves, and we comprehend them from the way in which they would affect us.

SECTION XVII.

QUALITIES OF MATTER: EXTENSION AND ENERGY.

Primary Qualities. — In all our sense-perceptions, even those simply of our bodies, there are qualities known. Some of these are called Primary. They are found in body, as Locke expresses it, in whatever state it be. They are so called also, because, as Reid says, our senses give us a direct knowledge of them. I doubt much whether we are able to determine with clearness and certainty what these are. Physical science will not pretend to fix on them absolutely. Metaphysics has no right to settle such a question. But it may be safely said that there are two such qualities: one of these is Extension and the other is Energy.

Extension is certainly an essential quality. Every form of matter possesses it. The intelligent mind directly perceives body as extended. By an easy process of abstraction we can separate the extension from the body as possessing other qualities and have the idea of extension or space. Hamilton evolves it from two catholic conditions of matter: "The occupying of space and being contained in space. Of these the former affords (A) trinal extension explicated again into (1) divisibility; (2) size containing under it density of gravity; (3) figure ; and (B) ultimate imcompressibility ; while the latter gives (A) mobility, and (B) situation."

Energy under certain forms is also an essential quality. Matter is known as affecting us and as resisting our action. True, it is only by a gathered experience that we know what forms physical energy takes, and find the nature, extent, and limits of the action, as, for instance, of gravitation and chemical affinity. But we seem in all our cognitions of body to know it as acting on us even as we know ourselves as acting on it. There is no form or state of body, solid, liquid, or gaseous, which does not possess this power which is exercised in our perception of it. It should always be acknowledged that matter may possess other essential attributes, as these may be known to other intelligences who penetrate into the nature of things. But these seem to be the only essential qualities known to us.

Organic Affections, called not very happily the *Secondary Qualities* of Matter. In regard to what secondary qualities are, such as smells, tastes, sounds, colors, there has been much controversy gendered of confusion, and many wrong inferences have been drawn. It is asked whether there is color in the rose, sound in the drum, odor in the violet, taste in the mutton. If we answer that there is, then it is shown conclusively that colors consist of vibrations, as do also sounds, and that tastes and smells are mere liquids and vapors affecting our palate and nostrils. But when we are driven to allow that there is no reality in these secondary qualities, it is argued that there may just be as little in the primary qualities, such as extension and resistance, which may be mere sensations of the organism or creations of the mind. The logical conclusion is idealism such as that of Berkeley.

The secondary qualities have an existence simply in our animated and sentient frame. Their office is to make known the state of our bodies. They do not reveal directly the properties of bodies beyond our organism; but they prompt us to inquire into the cause of the affections when we find them to consist of the mechanical or chemical properties of objects. It is thus that the sensation of heat or cold leads us to inquire into the state of the temperature, and that certain odors may send us out in search of malaria.

There is an ambiguity in the phrases, sounds, tastes, colors, heat, and the like. They may mean simply affections of the sense, nerves, or the bodily qualities which produce the affection. Thus "heat" may mean the frame under a certain sensation or a mode of motion. It is of importance that when we are using these phrases we understand and explain what we mean by them. When we speak of feeling heat we do not mean a mode of motion, which is in fact the cause of our feeling.

It will be found that in all our organic affections (as indeed in all physical action) there is a dual or plural cause; there is an organic susceptibility and an extraorganic agent; there are tastes, smells, and colors, but these are called into action by sapid bodies, by odors, or vibrations. These two, the organic and extra-organic, are so mixed in our apprehensions that we are apt to identify them. That smell we know is produced by a rose, and we regard the smell as in the rose. We can thus so far understand that peculiar combined sensation and perception as to color which has so puzzled metaphysicians. By the eye we perceive a surface, but there is always associated with it a retinal color in the rods and cones. It is only by a process of abstraction that we can think of (we cannot image) the color apart from the shape.

For philosophic purposes the all-important distinction is between the qualities perceived immediately in all bodies — these are the primary qualities; and the organic affections implying by inference an extra-organic cause — these are called the secondary qualities. It is to be distinctly understood that there is a reality in both. The reality in the secondary qualities is merely in the affected organism, and we are justified in maintaining that there is such a thing. The reality in the other is in body, and we hold that this really exists.

SECTION XVIII.

IDEAS GIVEN BY THE SENSES: EXTERNALITY, SPACE, AND ENERGY.

We shall discover as we advance that every one of the original mental powers gives us a special cognition or idea. We may notice here that Sense-Perception gives us I. EXTERNALITY. We perceive all material objects as out of, and independent of, the perceiving mind. This is associated with II. EXTENSION. We perceive things as extended by all the senses, not only as Locke thought by sight and touch, but by smell, taste, and hearing; by all these we know our affected organism as in a certain direction and so in space ; by taste and smell we know the palate and nostrils as affected, and by hearing, our ear as affected. III. We perceive body exercising ENERGY. We do so especially by the muscular sense; we find body resisting our locomotive energy. Perhaps we have some vague sense of energy by all the senses: the objects perceived seem to affect us. But the sense of power is specially given by our energy and the resistance to our energy. And then we soon learn by experience that our organic sensations are produced by extra-organic causes, that our sensations of light and heat are produced by vibrations. We are thus made to feel that every body is possessed of power in exercise or ready to be exercised.

These three primitive cognitions are the root of all our ideas regarding matter. As Kant would say, but in a different connection, " They render experience possible." It is of importance thus to note and to specify what is the precise knowledge given by the senses that we may see clearly and ever keep it before us, that they do not and cannot yield us all our ideas ; and that there are other and higher ideas as of self, of thinking, and moral good which must come from higher sources.

CHAPTER II.

By this power we know self in its present state as acting and being acted on.

SECTION I.

IT MAKES KNOWN SELF AS WELL AS THE ACTS OF SELF.

At the same time that we perceive by the senses we are conscious of ourselves as perceiving. These two exercises are, in many respects, like each other. In both we perceive an object. By the senses we perceive an object external and extended, this table or that chair. But in consciousness we also perceive an object: we perceive self in a certain state, as thinking or as feeling, as in joy or in grief. By the one we know the various properties of matter as they come under our notice; by the other we know the various states of self.

It is of importance to notice that in self-consciousness we come to have a knowledge of self in a particular state. According to D. Stewart and the Scottish school, we know only the qualities of things, and not the things themselves. The correct statement is that we know the thing as exercising a quality. According to Kant and his school, we know simply phenomena — that is, appearances, and not things. But there never can be an appearance without a thing appearing. In self-consciousness we know the thing, the very thing, as appearing or as presenting itself to us. We have as clear and certain proof of our knowing the object — that is, the thinking self — as we have that there is before us an appearance.

CONSCIOUSNESS ACCOMPANIES ALL MENTAL EXERCISES. — In this respect consciousness differs in its mode of exercise from the other powers of the mind. I am not every instant remembering, or judging, or willing, but at every waking moment of my existence I am conscious. When I perceive a material object, when I recollect an occurrence, when I draw an inference, when I am sorrowing or rejoicing, when I am wishing or willing, I am conscious that I do so. In short, consciousness seems inseparable from the exercise of all our faculties and to accompany every operation of the mind.

It was an opinion entertained by Leibnitz, and has been held by many since his time, that we are unconscious of many of our mental operations. They point to acts of mind which have left effects behind them, but of which we have not the dimmest recollection. We are sure that we must have issued a great many volitions in passing from one place to another, but after they are over we cannot recollect one of them. The question arises, How are we to account for such a phenomenon? I believe it can all be explained by the ordinary laws of mind, without our calling in such an anomalous principle as unconscious mental action. I hold that we were conscious of the acts at the time, but that they were not retained, as there was nothing to fix them in the memory.

The exercise of the mind when thus engaged is not unlike that of a man in a boat, looking over its edge into the lake below, thus described by Wordsworth: —

> " As one who hangs down, bending from the side
> Of a slow-moving boat upon the breast
> Of a still water, solacing himself
> With such discoveries as his eyes can make

> Beneath him in the bottom of the deep,
> Sees many beauteous sights, — weeds, fishes, flowers,
> Grots, pebbles, roots of trees, — and fancies more ;
> Yet often is perplexed, and cannot part *
> The shadow from the substance — rocks and sky,
> Mountains and clouds, reflected in the depth
> Of the clear flood — from things which there abide
> In their own dwelling ; now is crossed by gleam
> Of his own image, by a sunbeam now,
> And waving motion sent, he knows not whence,
> Impediments that make his task more sweet."

Every word of this description might analogically be applied to the reflex process of the human mind, as it observes its own thoughts and reasonings, its sentiments and emotions. At times there is a dimness in the view which we obtain of them, at times our vision is crossed by a gleam of our own image — that is, our observation of the act so far disturbs the act ; the party observing is discomposed by the knowledge of an eye fixed upon him ; or, to vary our image, the thought, when inspected, is so far modified by the inspection, as the very thought that a man is sitting for his portrait will so far affect the expression of his countenance. Still, as we thus inspect this deep, we shall see far more beauteous sights than weeds, fishes, flowers, grots, pebbles, roots of trees ; we shall see the workings of those thoughts which give to man all his greatness, of those sentiments which give to man all his excellence.

CONSCIOUSNESS AND PERSONAL IDENTITY. — Consciousness cannot be said to furnish our idea of, or belief in, our personal identity, for consciousness looks solely to the present, whereas in personal identity there is a comparison between the past and the present. But consciousness reveals self as present. When we remember the past, there is involved a memory of self as remembering. We are thus in a position to compare the two

— the present self known, and the past self remembered, — and we declare the self to be identical. Consciousness thus supplies the two main facts on which our judgment as to personal identity is pronounced. The self at present may be depressed and sad, the self remembered may have been buoyant and joyous, but we declare the two to be the same, and cannot be made to pronounce any other judgment.

It is not consciousness, as it has been sometimes asserted, that constitutes our personal identity. Consciousness merely makes it known, or rather makes known the facts on which our judgment rests. We are persons, and we have an identity of person whether we notice it or no. We are persons, and have an identity of person not because we are conscious of it, but we observe it because it exists.

Bishop Berkeley drove the doctrine that in sense-perception the mind does not perceive the external object, but an idea in the mind, to its legitimate consequences. He argued that if we do not perceive an extended world we have no reason to believe that there is any such thing. There has been a like error held in regard to consciousness, by which it is said we know merely phenomena in the sense of appearances (so Kant held), merely appearing thoughts and appearing feelings. Fichte did for this theory of Kant what Berkeley did for the theory of Locke. He followed it to conlusions from which the founder of the German school shrank. "The sum total is this: there is absolutely nothing permanent either without me or within me, but only an unceasing change. I know absolutely nothing of any existence, not even mine own. I myself know nothing, and am nothing. Images there are; they constitute all that apparently exists, and what they know of themselves is after the manner of images — images that pass and vanish without there being aught to witness their transition ; that consist, in fact, of the image of images without significance and without an aim. I myself am one of these images. All reality is converted into a marvellous dream without a life to dream of and without a mind to dream, into a dream made up only of a dream of itself. Perception is a dream ; thought, the source of all the existence and of all the reality which I imagine to myself, of my power, my destination, is the dream of that dream." I meet the ideal skepticism, or rather agnosticism, so far as it relates to the external world, by maintaining that, by the senses, not only do we perceive phenomena, we perceive appearances ; we perceive things appearing, not merely qualities, but qualities of self, of self in such or such a state. The conclusion to which we have come is that as by sense-perception we have a positive, though of course limited, knowledge of material objects, so by self-consciousness we have a like knowledge of self in its present action.

SECTION II.

SENSE-PERCEPTION AND SELF-CONSCIOUSNESS COMBINED.

We have been looking at these two faculties separately. Let us now look at them together. By the former we obtain a knowledge first of our own bodily frame. This we do by all the senses. We know our body as out of the thinking mind, and the organs as out of one another, and in a certain direction in reference to one another. We also know certain affections which we call tastes, odors, sounds, and colors. We know all matter as extended and as offering resistance first to our body, and then to other bodies. But at the same time that we are thus perceiving, or indeed exercising any other power, we know self, and this successively in its various moods or modes. It is the business of psychology to unfold these. These two do not constitute all our faculties, or even our highest or chief faculties, but they are the first exercised of all our powers, and furnish materials to all the others, which are therefore dependent on them.

SENSE-PERCEPTION AND SELF-CONSCIOUSNESS GIVE US KNOWLEDGE. — This proposition is laid down in opposition to the very common statement that the mind begins with impressions, or ideas, or presentations, or phenomena. The mind commences its intelligent act with the knowledge of things: by the senses of body, our own frame or things beyond; by the inner sense, of the conscious mind in its present state and exercise. These powers may, on this account, be called the simple cognitive, because they give knowledge in its simplest form.

Some would not allow that what is given us by these powers is knowledge. And no doubt it is not scientific or systematized knowledge. But still it is knowledge — a knowledge of existing things — not ἐπιστήμη, but γνῶσις; not Wissenschaft, but Kennen. The arranged knowledge requires a previous knowledge, which it arranges. The systems or theories of philosophy which do not begin with knowledge can never get it by any subsequent or subsidiary process, and so are landed, whether their defenders allow it or no, whether they wish it or no, in Nescience, which declares that man can know nothing; or in Nihilism, which affirms that there is nothing to be known; or in what is now called Agnosticism.

THIS PRIMITIVE KNOWLEDGE IS SINGULAR. — It

may consist of what 'is afterwards discovered to be a
number of objects, but it is regarded at the time as one
thing. The eye may have before it a widespread scene,
with divers objects of different colors and shapes, and
some of them farther removed than others, but it contemplates them as one surface. It is by a subsequent
process and by higher faculties than the senses that we
distinguish one part of the scene from another, this tree
from this hill, the animal from the ground on which it
walks. The same may be said of our knowledge by self-
consciousness. We are not conscious of a thought as distinguished from a sensation; we are conscious simply of
mind as thinking or as sentient, as one or other, or both,
without designating them or distinguishing them. This
knowledge is said to be "singular," as opposed to
"universal." It is of one object as it presents itself,
without or within us; this wall, or this feeling. It is
by a subsequent and a discursive process that out of the
singular we form the general. But the formation of the
universal always implies individual things, out of which
it is fashioned.

THIS PRIMITIVE KNOWLEDGE IS CONCRETE — that
is, it consists of objects as they present themselves, of
objects with their qualities, not of objects apart from
qualities, or of qualities apart from objects, but of objects
as exercising qualities. We may, by a subsequent process,
separate the things thus known, the substance from the
quality, or the quality from the substance, or one quality
from another. Having seen a house, we can think of its
walls, or its windows, or its door, or its roof. But this is
by a process of abstraction, and not by mere sense-perception. But in order to an abstract notion, there must
be a concrete apprehension. All the apprehensions given
by the senses and self-consciousness are concrete — that

is, of things grown together (from *concresco*), or of things seen together (from *concerno*).

The principles laid down in this and the preceding sections undermine that transcendental philosophy which supposes that the mind starts with such general or abstract ideas as space and time, infinity and eternity, supposed to be innate, on which ideas it would raise a huge but unstable system of speculative philosophy. It can be shown that all these ideas appear first in a singular and concrete form. It is sufficient, in the mean time, to remark that in sense-perception we have not space in the abstract, but body contained in space and occupying space.

SENSE-PERCEPTION AND SELF-CONSCIOUSNESS MAKE KNOWN THINGS AS HAVING BEING. — In every exercise of the senses, we know things, this organ of our body, or this ball in contact with it, as existing. It is the same in every operation of self-consciousness; we know self as planning, or purposing, or in some other exercise. Seldom, indeed, do we take the trouble of affirming that we ourselves exist, or that the objects before us exist. We assume it as a thing which we know, and which will be granted us. We are inclined to affirm only what may be denied, and this will not be denied, and it is superfluous, and might seem affected in us, to make any formal statement on the subject. It is implied in the exercise of our two primary capacities that the things they look at have Being.

"But what can be said of Being? Verily, little can be said of it. The mistake of metaphysicians lies in saying too much. They have made assertions which have, and can have, no meaning, and landed themselves in self-created mysteries or in contradictions. So little can be affirmed of Being, not because of the complexity of

the idea, but because of its simplicity; we can find nothing simpler into which to resolve it.   We have come to
ultimate truth, and there is really no deeper foundation
on which to rest it.   There is no light behind in which
to show it in vivid outline.

   " In the concrete every one has the cognition of Being,
just as every man has a skeleton in his frame.   But the
common mind is apt to turn away from the abstract idea,
as it does from an anatomical preparation; or rather, it
feels as if such attenuated notions belong to the regions
of ghosts, where

> " ' Entity and quiddity,
> The ghosts of defunct bodies, fly.'

   " All that the metaphysician can do is to appeal to the
perception which all men form, to separate this from the
others with which it is joined, and make it stand out
singly and simply, that it may shine and be seen in its
own light, and with this the mind will be satisfied : —

> " ' Who thinks of asking if the sun is light,
> Observing that it lightens ? '

Those who attempt anything more, and to peer into the
object, will find that the light — like that of the sun —
darkens as they gaze upon it.   ' When I burned in de-
sire to question them further, they made themselves —
air, into which they vanished.' "

. The Eleatics, who flourished five and six hundred
years before Christ, made much of Being το ὄν, and were
followed by the Greek philosophers generally.   I do not
believe that they attached too much importance to this
idea.   That there are existing things is the fundamental
position in metaphysics.   It is a fact to be assumed, and
no attempt should be made to prove it.   Any professed
proof will turn out to be delusive, as we cannot find

anything simpler or more certain by which to establish
it. The fault of the Greek philosophers, and especially
of the Eleatics, consisted in making affirmations about
Being which have no meaning. All that we can say of
Being is that it is Being.

THEY MAKE KNOWN THINGS AS EXERCISING PO-
TENCY. — It might be maintained that through all the
senses we know bodily objects as exercising power over
us. We know tastes and smells, and colors and sounds,
as influencing us, and producing a change in us. We
certainly know objects as resisting our muscular energy.
It is equally certain, some would represent it as more
certain, that we know the will and other mental faculties
as exercising power over the body and over states of the
mind. Potency is thus an element in all primary cognitions. Everything we know we know as exercising
power on us or on some other object.

(1.) It is clear that if we do not know power intuitively we can never know it by any derivative or discursive process. But consciousness being our witness,
we have an idea of power quite as certainly as we have
of extension or of thinking.

(2.) While we obtain in this way our knowledge of
things within and without us as exercising power, it is
only by the gathered experience that we are able to determine what is the precise nature of that power, what
its laws and its bounds. All that we know directly of the
power of matter by the senses is very limited. We know
odors, and tastes, and colors as producing a sensitive
affection in us. What these are, and what their properties in other respects, we have to learn by a process of
observation ; and we discover that odors affect us only
when in a state of vapor, tastes only when the bodies are
liquid and that sounds and colors are made known by

undulations. By the muscular sense we know bodies simply as resisting our energy, and we have to go to physical science to determine what are the laws of energy generally. It is the same with the power exercised by any of our mental capacities. We know that there is power to produce an effect in every operation of the mind, but it is the office of psychological science to determine the rules and limits of the faculties.

THEY MAKE KNOWN THINGS AS HAVING INDEPEND- ENCE; THAT IS, AS EXISTING INDEPENDENT OF THE CONTEMPLATIVE MIND. — The thing does not exist merely because the mind contemplates it. The mind contemplates it because it exists. It does not begin to exist when I begin to notice it. Nor does it cease to exist because we have ceased to observe it. We have all this involved in the knowledge conveyed both by the outward and inward senses. This does not imply that the thing has any absolute independence, that it is inde- pendent of God. All that is meant is that it exists in- dependent of the mind taking notice of it.

By laying down this position we are delivered from a position taken up by many in the present day, and which lands them first in confusion, and in the end in skepti- cism. Taking advantage of the ambiguity in the use of the phrases object and subject, they tell us that object always involves subject, and subject object, and that in fact our knowledge, if knowledge it can be called, is made up of two factors which cannot be separated. The result is that we cannot tell what any one external object is, for it is mixed up with the subject mind, which gives it in a certain form and a color. On the other hand, we can scarcely ascertain what the subject mind is, it is so dependent on the objects which call it into exercise The result of the whole is a growing feeling of doubt as to the reality of things. Even when they are acknowledged to be real, it is supposed to be impossible to determine the precise nature of the things supposed to be real. Now, this subtle metaphysical error is to be met by affirming that the thing is contemplated as it is, and that the subject mind is so constituted as to be able to cognize it. If asked for proof of all this, the reply is that we have the same evidence of the mind contemplating the thing as it is that we have of its contemplating the thing. It should be the business of metaphysics not to confound the subject and object, but to point out clearly the distinction between them.

SECTION III.

SUBSTANCE.

We have seen that both body and mind are known by us through the senses as possessing Being, Independence, and Potency. Whatever possesses these may be regarded as a substance. Some would have us say, fourthly, having independent existence, or independent of any creature. But the difficulty is to determine what constitutes independent existence. All things are dependent on God, and seem more or less dependent on other things. Still there is vague truth in the statement. These marks give a definite meaning to the phrase. We see that there are two substances known to us — mind and body.

Hamilton says that substance may be regarded as derived from one or other of two words: from *substo*, to stand under; or from *subsistc*, to subsist of itself. Descartes defined substance as that which subsists of itself. Spinoza gave a more complex definition: " By substance I understand that which is in itself and conceived by itself; that is to say, that of which the concept can be formed without needing the concept of any other thing." Locke understood substance as something that stands under. It is evident that substance, thus understood, must come in very awkwardly under a system which derives all our ideas from sensation and reflection, as it cannot be derived from either of these sources. He does not deny the existence of substance, but he represents it as something unknown and unknowable. Most of his followers contrived some way or other to get rid of this unknown thing as being something superfluous, and of the existence of which we have no proof. In the text, substance is represented as a thing known and involved in our intuitive knowledge both of body and mind.

BODY IS A SUBSTANCE. — It is so, according to our definition. We know it as existing, as existing independent of our cognition of it, and as exercising power.

Locke, we have seen, represented substance as an unknown support of things. Berkeley showed that there was no evidence of body having any such support. He did not deny the existence of matter, but he denied that it was a substance. We meet Berkeley not by standing up for a support or substratum unknown and unknowable, but by maintaining that we actually know body as having an abiding existence.

MIND IS A SUBSTANCE. — We make this affirmation on the same ground as we maintain that body is a substance. In every act of consciousness we know it as exerting and exercising power, and this independent of our taking any observation of it.

As Berkeley denied that body is a substance, so Hume denied on much the same grounds that mind is a substance. He represented it as a mere series of percep

tions, with a unity given to it by the imagination. Now
we meet this by showing that in every act of consciousness we know self as existing and exercising potency of
some kind.

MIND AND BODY ARE DIFFERENT SUBSTANCES. —
In this respect they are both alike: that they are substances. As such, they have the three points of affinity
so often mentioned, and they may have many others.
There may be correlations of an important kind between
their various properties, but they are known to us as
different. In particular, first they are known to us by
different organs: the one by the senses, the other by
self-consciousness. Then, secondly, they are known to
us as possessing very different attributes: the one is
known as extended and resisting, the other as thinking,
musing, resolving. These differences entitle us to regard them as different substances.

Descartes separated mind and matter so entirely that
the one could hold no communication with the other
except, as Malebranche brought out more fully, through
an interposed divine action acting as an occasional cause.
Proceeding on the same principle that mind and matter
could not act on each other, Leibnitz brought in his
doctrine of Preëstablished Harmony, according to which
they act in unison, not by reciprocal action, but by an
order established in each, whereby, like two clocks, they
correspond the one to the other. But there is no need
of resorting to any such far-fetched hypotheses. We may
suppose that the two act and react on each other, according to laws not yet determined. An action goes along a
sensor nerve to the sensorium, and is thence transmitted
to the periphery of the brain and to the cells there,
where it calls forth a mental power, with which it co-
operates, and becomes a perception of an external object,

say a rose. The rose may, according to purely mental laws, give rise, by an association to an entirely different idea, say to a lily, and we may then compare the rose and the lily. The law of the conservation of physical force must regulate all the action as far as the cells in the circumference of the brain. When the action becomes purely mental, as in all recollections, judgments, imaginations, moral sentiments, and volitions, there is no reason to believe that the doctrine of the conservation of energy has any direct place. Still it is conceivable that even in purely mental acts there may be a laid-up physical energy, which goes out in brain action. All this may be admitted without giving any countenance to materialism. It has all along been allowed that, as man is constituted, mind and body have a very intimate connection, and this may be the way in which this connection is kept up. But we need a great many careful observations and experiments before we can determine the precise relation of physical and mental potency.

SECTION IV.

LOCKE'S THEORY AS TO THE ORIGIN OF OUR IDEAS.

Locke gets the materials of all our ideas from Sensation and Reflection. By sensation, he means the same as the Greeks did by αἴσθησις, and as we do by sense-perception; and by reflection, much the same as we do by self-consciousness. Upon the materials so supplied certain faculties work, such as Perception, Retention, and thus fashion all our ideas. This theory will require to be criticised as we advance, and it will be shown that there are ideas such as that of moral good and evil, which cannot thus be obtained. But, meantime, let it be remarked that by these two inlets we get a great many of our ideas; by the senses of bodies as external to us, as extended and resisting our energy and resisting one another, and by self-consciousness of the mind in its various states, say perceiving, remembering, imagining, judging, discerning between good and evil, under emotion, or as resolving.

The word reflection might now be applied to the more special notice which the mind takes of itself and its operations. In this there is an exercise of will joining on to self-consciousness; it is a voluntary consciousness. It is mainly by this power that the science of Psychology is constructed. We observe the operations of the mind as they pass, and thus are enabled to analyze, to classify, and arrange them.

SECTION V.

TRAINING TO HABITS OF REFLECTION.

Man is naturally inclined to look out of himself before he looks within. There is a propriety in this. The mind must have materials of thought before it thinks. But it is of importance that we be trained to bend back our attention to and notice what is passing in our minds, and thus know ourselves. We shall be led into great mistakes if we do not from time to time look into our inward state and search our motives. This, I admit, may be carried too far. There may be too much of self-consciousness; no, not too much, but a misdirected self-inspection. Instead of allowing the plant to grow under the air and sunshine provided for it, we may be injuring it by ever searching into its roots to find whether it is growing. Still, reflection, which always includes inspection, is one of the peculiar properties of humanity, distinguishing man from the brutes, and should be called forth in the

opening years of manhood and womanhood, and continued through life. Spontaneous thought comes forth
first, constituting what is called "first thoughts;" but
reflective thought should come after to detect error, to
cast off the mistakes associated with the truth, and secure
certainty. We should not be satisfied with things as they
appear nor with first impressions or first thoughts, nor
with the opinions we have formed in the past; we must
acquire and train a habit of self-examination, and make
them all pass in review before us.

BOOK SECOND.

THEY are so called because they produce and present once more, and it may be again and again, what has been previously before the mind. Some of them are farther representative, inasmuch as the ideas raised up by them stand for absent objects: thus the memory brings up an object or event once before the mind, but not now present. This can scarcely be said of them all, as for instance the imagination, in which there is no other object than the image itself.

I have seen Mont Blanc. Having done so, I retain it in such a way as to be able to recall it. It comes up from time to time in the shape of an image according to the laws of association. It is recognized as having been before my mind in time past. I can put it into new forms and dispositions. I can think and speak of it by means of the name which has been given it. In such an exercise we have the mind exercising six different capacities; these we call

I. THE RETENTIVE.	IV. THE RECOGNITIVE.
II. THE RECALLING OR	V. THE COMPOSITIVE.
PHANTASY.	VI. THE SYMBOLIC.
III. THE ASSOCIATIVE.	

It has been shown (Introd., Sect. IV.) that the mind or self possesses power, or rather powers. I am now seek-

ing to unfold the various faculties. But it is to be understood that these faculties are not separate personali-
ties or things. They are simply modes or activities of
the one self. Thus Sense-Perception is the mind perceiving external objects, and Self-Consciousness is the
mind perceiving self. The same remark may be made
as to the other powers. Thus the Memory is merely the
mind remembering past experiences ; the Conscience,
the mind discerning good and evil ; the Will, the mind
choosing. This general truth holds true of all the faculties ; it should be remembered, but need not be repeated
under each head. It is also to be kept in mind that the
powers do not act independently of, but rather with, each
other. The Phantasy and Association proceed on the
Retentive power. We shall see that in the Memory and
in the Imagination there are several powers involved,
and that the one supplies materials to the other. By
taking these views we avoid the objections of Herbart
and the metaphysicians of the school of Leipsic, who
complain of the way in which the mind is mangled and
the parts are separated by psychologists.

CHAPTER I.

THERE is a difference between the state of our minds before we have observed an occurrence and after we have observed it. There is a difference between one who has noticed an event, or passed through an experience, and one who has not. Having been at the London exhibition of 1851, I have something continuing with me which I could not have had unless I had been there. Having once passed through a period of severe illness, having passed through the disruption of the Church of Scotland in 1843, I have a series of impressions and lessons not possessed by those who have not had such an experience. Whatever has passed under consciousness may be retained; it always produces some effect which may remain. It is so retained that it can be recalled according to certain laws of association.

We cannot say much more about the conservative power, as Hamilton calls it. In what state is an idea, say of London or Paris, when it is not immediately under the consciousness? Is it dead, or simply dormant? It is certainly not altogether defunct, for it can be wakened. We have something analogous (though not identical) in the energy potential as distinguished from the energy real or kinetic in physical operation. The energy which came from the sun in the geological age of the coal measures is laid up in the coal, and comes out in certain

circumstances in heat to warm our bodies or drive our
steam engines. Having passed through a conscious experience, the mind has the acquired capacity of calling it
up. It is actually recalled when there is in the mind
an idea associated with it. The retention depends

First, On the state of the brain, more especially on cells
in the gray matter on the periphery of the brain. Every
one has felt that in certain states of the brain we have
difficulty in remembering anything. The ardent student,
the anxious business man, may so exhaust his cerebral
force that nothing will be retained in his mind. In such
cases perfect rest, particularly "balmy sleep, is nature's
sweet restorer." From probably much the same causes,
we find that when we are engrossed with any one care
or distracted by several things we are apt to forget the
extraneous things which have passed before us momentarily; a piece of news in which we are not particularly
interested, given us at a time when we were absorbed
with other things, may never come up again.

We have come into a border country where there is a
constant warfare raging, and it is difficult to determine
the exact bounding line between mind and body. This
admission does not go to establish materialism. Every-
body grants that mind and body are intimately connected, and we have simply come upon one of the points
of connection. No intellectual faculty of the mind is so
dependent on the brain as the memory, and retention is
one of the conditions, or rather one of the concurring
agencies, in memory. There are some positions which
can be defended. Every idea, every feeling, is thought
to tend to produce an effect on the periphery of the
brain, and probably to give a particular disposition or
set to the cells in that region. It may be maintained
that the concurrent action of the part of the brain

affected seems to be necessary to our recollection of an occurrence. When the idea or feeling produces little or no effect on the brain there may be no recollection, or only a very dim one. When there is a lesion or a disease in the brain, or in certain parts of it, we are apt to lose our memories, or have them deranged.

Secondly, Retention depends on the mental force in the original feeling or idea. This second condition may be connected with the first. The strong or lively thought produces a deeper impression on the brain, which aids the remembrance of it. But the two essentially differ. The profundity of the thought or the power of the sentiment is not caused by the organism, say the discoveries of science, or the affection of a mother. We must all have noticed that events which have not interested us, or to which we have given no attention, are apt to pass away speedily from the memory, whereas others, which have exercised our understanding, or called forth emotion, are remembered for years or our whole lives. It is no matter what the sort of mental power directed towards an event be — whether it be the intellect, the affections, or the will — it tends to keep it ready to be called up. On the other hand, when no special mental power is exerted the occurrence may never come up again. This is a subject worthy of being prosecuted and illustrated, and opens to us many interesting and instructive views of the operatiohs of the mind. But it may be expediently deferred till we come to speak of the secondary laws of association — those that modify the primary and make them take a particular direction.

The laws now announced, and to be afterwards more fully expounded, may help to explain what are called unconscious mental operations; that is, operations which have passed in the mind, but of which we are not conscious. There are, undoubtedly, mental exercises which are not recalled in ordinary circumstances. There are acts of the will implied, and I believe also of the understanding, in every step which a man takes in walking towards a particular place. The foot will not move without a volition of the mind, and there is thought implied in its carrying him towards an intended place. Yet at the end of his walk he may not remember one of the acts of his will or judgment. It is not just correct to call these unconscious acts. He may have been conscious of each of them at the time, and if there was anything to call his attention to them — say his taking a false step — he would have felt that he had been conscious of them and he would have remembered them. But there was nothing in the ordinary steps taken to make him notice them, and so they passed away. There was a momentary consciousness, but there is no memory of them. I do not agree with the theory of those who ascribe the creations of genius — say Shakespeare's Hamlet, or Milton's Satan, or Goethe's Faust — to unconscious mental action. True, these men might not be able or care to analyze like a metaphysician the processes that passed in their minds; but there was a cognizance of them at the moment in their concrete state, and there may have been a joy in them. There may not have been a consciousness of them in the sense of rolling them as a sweet morsel under the tongue. They passed through the minds as the fresh wind passed, by breathing, through the bodies; but they were not detained to cherish a feeling of self-complacency, and the poets passed on to some new thought or emotion.

The question has often been started, Do we remember everything and forget nothing? I am not sure that we can certainly decide this question. On the one hand,

there have been thoughts and feelings in our minds
which never have returned. On the other hand, there
are experiences which start up like ghosts from the grave
where we imagined we had buried them. We have all
of us had memories coming up unexpectedly of friends,
of incidents, that had not been thought of for long years,
and that now appear to give us joy or reproach us. But
I cannot believe that every one of the hundreds of sensations, of fancies, of opinions, of fears and hopes, which
pass through our minds in a few minutes, is capable of
being reproduced. It is a happy thing when thus our
trivial thoughts pass into oblivion ; otherwise our minds
would be filled with innumerable details and become as
trifling as these are. Those that come up unexpectedly
do so because they have left a deep impression at the
first, and they awake because stirred up by some correlated present thought.

We have curious instances recorded of persons who
have lost certain recollections, certain kinds of recollections, while they retain others, their minds all the while
being otherwise entire. It has occurred to me that we
may be able so far to account for these phenomena. It
is allowed on all hands that many of the operations of
the mind are dependent on cerebral coöperation, without which they would cease, or be carried on with difficulty. There are physiologists who allot a special
locality to each of the senses in the brain. It appears to
me that the concurrent action of the sense centres, or at
least of the brain, may often or always be necessary to
the recalling of the scenes perceived. When there is an
imperfection or a lesion in any of the sense centres, it
may be difficult or impossible to produce a phantasm of
the object, or it may be faint or disfigured. I have
heard of persons who had not lost their eyesight, but

owing apparently to a disease in the brain had lost the power of recalling the visible scenes they had witnessed. It is well known that the remembrance of forms and colors by persons who have become blind is apt in time to become dim. The same may be true of the other senses. When the organs of taste and smell, supposed by Ferrier to be in the back of the head, are diseased or out of order, the reproduction of the corresponding sensations may be indistinct. Tunes cannot be recalled, it may be presumed, when the organs of Corti are not in healthy working order.

It is generally believed that the fore part of the brain is more specially connected with intellectual action; and disease there will be apt to affect our recollection of all operations requiring thought, such as scientific truths. Perhaps the cerebral lobes in the fore parts are more particularly the centres of motion and our ideas of motion; and when there is a lesion in certain parts we may find difficulty, as some do, in imaging movements.

It is now acknowledged by almost all that M. Broca has established that there is some connection between the third convolution of the left side of the brain and the power of using language. When there is disorganization in that part there is experienced a difficulty in recalling words, especially names, or in making an appropriate use of them.

CHAPTER II.

As long as an object is merely retained it is not before the consciousness, and in fact may never be so. But it may come into consciousness according to laws of association to be unfolded in the next chapter.

Every man, woman, and child has a chamber where he or she has laid up a store of images or photographs of the objects which have been perceived. It may be interesting to take a look into it and inspect its contents, which will be found to be very curious. Every man has his own chamber of imagery with its separate furniture, grave or gay. It is the place of figures and fancies.

I call the power which reproduces in old or in new forms our past experiences the Phantasy, a phrase employed by Aristotle to denote one of the faculties of the mind, and which was used in the English tongue down to the beginning of the last century, when it was abbreviated into Fancy, with a more confined meaning. The product may be called the Phantasm — always to be distinguished from the phantom, in which the object is imaginary. Phantasy is a good phrase to designate the remembrance or imaging of a single object, say a lily, as distinguished from a general idea, such as the class lily. The faculty may also be called the Imaging or Pictorial power, only there is no image or picture except when the reproduction is of an object perceived by the sense

of sight — the other senses, however, being also capable
of reviving what has passed before us.   It is the mind's
eye of Shakespeare : " In my mind's eye, Horatio."

All these phrases are figurative, always implying and
pointing to a reality.   We talk of an image, a likeness,
a representation, an idea.   In what sense ?   So far as
the sense of sight is concerned, there is an image on the
retina of the eye.   But this is so situated that it is not
seen naturally ; in fact, it has been discovered by science.
The object is perceived upright, but it is inverted in the
eye.   Then, so far as the other senses are concerned,
there is no image, properly speaking.   There is merely
an affection of the organ — of the ear, the touch, the
palate, the nostrils.   Speaking rigidly, there is no image
of a taste or a sound.   Even so far as vision is concerned,
the image on the retina cannot be said to be perceived
by the mind.   It is merely an affection of the organism,
of such a kind that it becomes the fitting means by
which the exact form and color of the object are known ;
just — and not otherwise — as an ear makes known the
sounds emitted.   In respect of an image, there can be no
such thing in the brain in regard to any of the senses.
In all the senses there is an affection not only of the
physical part of the senses proper, but of the brain ; but
this does not take the shape of a form of any kind.   If
there is no figure in the brain, still less can there be in
the mind.   A figure is an extended material thing.   The
figure of a tree is no more in the mind than the tree is.
In all the senses the perception is simply a knowledge of
an object under a certain aspect, say as having a form or
odor.   In this sense only is an idea the representation of
an object.   There is really no likeness between gold as
out of the mind and the idea of gold in the mind.   There
is a correspondence between the two, but no identity.

In fact, this imaging power is merely one of the factors in the memory. In memory there is a recognition of an object or event as having been before us in time past. But in the mere imaging there is no such recognition and no reference to time. We may have a phantasm of a flower without any belief as to where or when we saw it, or indeed as to whether we ever saw it. But in all proper memory there is an image or phantasm, dull or vivid, representing the object or event recognized.

It has to be added that the mind has the power of forming imaginary figures. These are compositions constructed by the mind out of realities experienced. We have now, not memory, but imagination. Our imaginations, as every one knows, are often more lively than our recollections. The mind delights to form such pictures, and it is the office of the poet and novelist to raise them up by the presentations they furnish.

First, We can thus reproduce the material got by any of the senses. We remember tastes of salt, of sugar, of jelly, of apples, of oranges, and hundreds of other things that are sour or sweet, or do otherwise powerfully affect our palate pleasantly or unpleasantly. These recollections are not especially inspiring or poetical, but are cherished by gourmands, who feel as it were the taste in their mouth of the food they relish. We can recall the sensation produced by odors, say from roses, lilies, and violets, or from assafœtida, swamps, and malarial pools. Some of these are of an ethereal nature, and have a place allowed them in poetry. We can call up a thousand kinds of sounds, as the voices of our friends, the sighings of the breeze or stream, the barking of the dog, the mewing of the cat, the bellowing of the bull, the lowing of cattle, the chirp or the song of birds — say of the thrush or nightingale, the screech of the eagle, the

7

rasping of the file, the mower whetting his scythe, the roar of the storm, the lashing of the wave on the shore, the rolling of the thunder, the crash of the avalanche. People endowed with a musical ear can recall tunes, and are prompted to repeat them, and some are constantly hearing musical airs.

> " Music, when soft voices die,
> Vibrates in the memory ;
> Odors, when sweet violets sicken,
> Live within the sense they quicken."

There are touches which we easily remember — of softness or smoothness, say of satin or of a smooth skin, or of the prickliness of a brier or thorn. The child retains forever the memory of a mother's kiss. But we get our most vivid and varied memories from the sense of sight. We delight to remember colors, say of a flower or a piece of dress, of the morning and evening sky. We image certain forms, as of the persons and faces of our friends, of noble trees, of well-proportioned buildings, of mountains. All that is picturesque, that is picture-like, that is with a well-defined shape, as steeples, cliffs, precipices, leave a photograph of themselves on our souls. The artist uses many of these in his paintings, in his portraits, and in his landscapes. The poet turns them to all sorts of uses in pleasing, in exciting and elevating the mind.

This imaging power helps greatly to enliven our existence. We call up an incident of our childhood. We remember the day on which we were first sent to school, and how we set out from our parents' roof with strangely mingled feelings of confidence and timidity. As we bring back the scene, mark how everything appears with a pictorial power. We have a vivid picture, it may be, of the road we travelled; we see, as it were, the school-house, within and without; we hear the master addressing us, and the remarks which the children passed upon us. Or, more pleasant still, we remember a holiday trip in the company of genial companions or kind relatives to a place interesting in itself or by its associations; or the visit we paid to the house of a good friend, who had a thousand contrivances to please and entertain us. How vivid at this moment the picture before us of the incidents of the journey; of the little misfortunes that befell us; of the amusements provided for us; of the persons, .the countenances, the smiles, the voices and words, of those who joined us in our mirth or ministered to our gratification. We not only recollect the events: we, as it were, perceive them before us; the imaging is an essential element of our remembrance. Wordsworth is painting from the life when he speaks of

> " Those recollected hours that have the charm
> Of visionary things; those lovely forms
> And sweet sensations that throw back our life,
> And almost make remotest infancy
> A visible scene on which the sun is shining.

Or possibly there may be scenes which have imprinted themselves more deeply upon our minds, — which have, as it were, burned their image into our souls. Let us throw back our mind upon the time when death first intruded into our dwelling. We remember ourselves standing by the dying bed of a father, and then we recall how a few days after we saw the corpse put into the coffin and then borne away to the grave. How terribly distinct and startling do these scenes stand before us at this instant! We see that pallid countenance looking forth from the couch upon us; we hear that voice becoming feebler and still feebler; and then we feel as if we were looking at that fixed form which the countenance took when the spirit had fled; we follow the long funeral as it winds away to the place of the dead, and we hear the earth falling on the coffin as the dust is committed to its kindred dust.

Secondly, It should be specially noticed that not only are we able to represent these sensible scenes: we are further *able to picture the thoughts and feelings which passed through our minds as we mingled in them.*, Not only do we remember the road along which we travelled and the building which we entered : we can bring up the feelings with which we set out from our parents' house, and those with which we passed into the school. Not only do we recollect the amusements which so interested us, but the feelings of interest with which we engaged in them. Not only do we picture the chamber in which a father breathed his last : we can call up the mingled emotions of anxiety, of hope or fear, with which we watched by his dying bed, and the grief which overwhelmed us as we realized the loss we had suffered. We bring up the feelings which chased each other as we sat by his corpse, or when we returned to our home and felt all to be so blank and melancholy. ·

We can thus live our mental experiences over again : the efforts we made to acquire a branch of knowledge, a new language, or a new science, and how we found the process to be irksome or stimulating; what we felt in our failures or our successes, in our fights and in our triumphs, in our friendships and in our enmities, in our temptations yielded to and our temptations resisted. As we survey the past, we can remember the gratitude we felt on kindness shown us, the sorrow that overwhelmed us on the death of a friend, the bitterness of the disap pointment when our best hopes were frustrated, when one we trusted betrayed us, and the pang that shot through us when we found that we had committed an unworthy deed. We are obliged to use metaphorical language in describing these recollections. We speak of our being able to image or picture to ourselves the outward incidents and the inward feelings, and we thus set forth an important truth.

True, we cannot give these mental states a sensible figure. The reason is obvious. They had no visible or tangible form when we first experienced them, and the memory, in reproducing them, will represent them as they first presented themselves. This circumstance, I may add in passing, furnishes an argument of some little force in favor of the immateriality of the soul. In our primary knowledge and in our subsequent recollection of bodies we have a sensible image. But in our consciousness of our mental states and in our recalling them, we do not, and indeed cannot, so represent them. We give a bodily shape to the school at which we learned our tasks, to the persons and countenances of our early associates, but we cannot give a form or local habitation to our remembered cogitations and sentiments, which live in a higher sphere.

It is conceivable that the memory might have been as correct as it is of matters of fact without having any pictorial power. In fact, the majority of our memories must be of this character. It is well it should be so, for otherwise excitement would waste our life, and keep us from the performance of many commonplace but important duties. But that is a most benignant endowment whereby we can image absent objects and past events, lay them up in " chambers of imagery," and make them pass as in a panorama before us. We can thus have a series of paintings of all the scenes in which we have mingled, a set of portraits of the friends with whom we

had sweet intercourse, and we can view them as Cowper
did his mother's portrait: —

> " Faithful remembrancer of one so dear;
> And while that face renews my filial grief
> Fancy shall weave a charm for my relief,
> Shall steep me in Elysian reverie.
> A momentary dream that thou art she,
> By contemplation's help not sought in vain,
> I seem to have lived my childhood o'er again — '
> To have renewed the joys that once were mine,
> Without the sin of violating thine.
> And while the wings of fancy still are free,
> And I can view this mimic show of thee,
> Time has but half succeeded in his theft —
> Thyself removed, thy power to soothe me left."

This imaging power, as it enlivens the mind, also tends
to give vividness to its productions in words and writings. He is an interesting companion who, having laid
up a store of pictures, is ever bringing them out in his
conversation. Travellers and biographers instruct us
best when they are able to give us a word-painting of
the scene and of the man or woman. History is vastly
more attractive when it gives the event with its concomitants — say the battle with the field on which it was
fought. Our pictorial writers are generally the most
popular. In the mediæval ages they illuminated the
manuscripts to attract and delight the eye. In our day,
books in almost every department of literature are illustrated. This power has a still more important function.
Nothing tends more to degrade the mind and sink it in
the mire than low and sensual images. On the other
hand, images of duty, of self-sacrifice, of courage, of
honor, of beauty, of love, elevate and ennoble the soul.

Some of the phantasms are much more vivid than
others. They differ also in the case of different individuals, and of the same individual at different times or

in different states of his body. It is a curious question
what can be the cause of this difference. Without professing to exhaust the subject we may specify some circumstances which undoubtedly have an influence on the
vividness of the picture.

1. There is the original vividness of the sensation, depending primarily on the sensitiveness of the organ, but
under this also upon the nature of the object perceived.
The senses evidently differ in this respect. The most
lively is the sense of sight. The forms and colors originally made known by it may come up almost with the
distinctness of the realities. The mental representation
(we can scarcely call it picture) of sounds is often very
intense, especially in the case of those who have a musical ear, but also when the impression on the ear is strong
or vehement, — made, for instance, by the bursting of a
cannon. Tastes and odors may also be recalled with
less impressiveness, as also touches and feelings in our
nerves. There are times when our sensations of shapes,
colors, and sounds are very intense, and in these cases
they are apt to be reproduced with greater vividness.
There are scenes of gorgeous coloring, there are picturesque figures, such as horrid precipices; there are sounds
such as those of a falling rock, of thunder, or of an avalanche, which we can never forget. Some persons are
evidently more susceptible of intense impressions than
others, and in these cases the images are apt to be more
vivid, and these may be embodied in paintings, in statues, or in word-painting in prose or poetry.

2. The formation of the image is dependent on the
state of the brain. It is believed that even in our sense-
perceptions there is brain action. It seems to be established that the third convolution of the left side of the
cerebrum is the organ of the symbolic power, or of lan-

guage. Some eminent men, such as Hitzig and Fritsch and Ferrier, maintain that each sense has a separate location in the brain ; others deny this. Without entering into this discussion, it is allowed that brain action is necessary to sense action. The whole eye might be perfect, and yet there is no vision if there be a lesion in certain parts of the brain. Not only so, but brain action is required in order to the reproduction of our sense-perceptions. Now it is highly probable that the same part of the brain acting in the perception is necessary in order to its reproduction. When there is a lesion of a certain part of the brain it may not be possible to form an image of the object. In all cases the vividness of the image may depend on the health and susceptibility of the brain matter.

It is well known that persons may lose certain of their recollections while they retain others. The defect seems to arise from a lesion of the brain. We have the record of persons losing the power of picturing forms, while their memory was good in all other respects. We have more frequent instances of people losing their power of using languages or particular languages. This is the disease of aphasia, arising from a derangement in the organ of language. There are cases of persons losing a portion of their knowledge for a time and then recovering it ; perhaps losing it suddenly, and recovering it as suddenly. In all such cases it looks as if, in acquiring the original knowledge, there is a certain state of the brain produced, say by a certain disposition of the molecules, probably in the gray matter in the periphery of the brain. Where there is an effacement or derangement of this matter in the brain the knowledge cannot be recalled. Sometimes the disorganization is only for a time, and when it is cured the mental power is ready to act.

3. There is the mental force particularly of the attention directed to the scenes as they first passed before us. We were interested in them, we turned them round and round, we viewed them under various aspects, and having been so encouraged and fondled, they are apt to visit us again and again, and put on their best expression. The painter has to study the features of landscapes and the countenances and attitudes of men and women to give us correct figures on his canvas. Under this view, the capacity of bringing up images is more within our power than we might at first imagine.

SECTION II.

CHAMBERS OF IMAGERY.

Following the plan of Professor Galton in his "Questions upon the Visualizing and Allied Faculties," Professor Osborne and myself issued certain queries to the students of Princeton and Vassar (Female) Colleges. The answers are very curious, and I may detail some of them.

The Phantasy is exercised most vividly in regard to the sense of sight. The following are the answers of various persons : —

(1) I can recall the features of some exceedingly well-known persons, as of my own family; (2) It is hard for me to image faces with great distinctness of detail; (3) I can recall comparative strangers with more ease than near relatives ; (4) I can recall the features of many persons, of almost any one, better than of my friends and relatives ; (5) I can recall the features of all whom I have ever known intimately, except my mother ; (6) I frequently recall faces with vividness, *but not at will ;* (7) I can recall the features of males better than of females; (8) I can only recall the features of those who have been lately seen ; (9) There are a few persons very well known to me whose features I absolutely cannot recall, and it is very annoying ; (10) I can recall readily persons, friends, and relations ; (11) I can recall all quite distinctly, but those with whom I am associating every day with more distinctness than others, as my classmates at college better than my friends at home.

The images formed in childhood are with most persons clearer, brighter, and more numerous than those of later years. Among twenty-eight students three believe that their powers of imagery have improved, thirteen say that they have not varied, twelve say that they have diminished. This is due in many cases to disuse, for there can be no doubt that the elaborate imagery of some older minds is far more wonderful than anything found among children. Children's images, apart from the natural strength of their phantasy, are vivid because they see form, color, and outline dissociated from any distracting ideas which would enter the mind of an adult. A child looks at a pony, engrossed with its external characters, rough coat, long mane, and so on, without thought of price, age, or disposition. This concentration and simplicity of the mental concept affects the memory as sharp focussing affects a sensitive plate. The earliest images recalled from childhood are amusingly trifling ; they are often of objects which touched the childish vanity, such as the first long trousers or new blue dress, the first day at school.

The following experience of a young man, now a physician (Dr. Loyd), is full of instruction : —

" A year or two ago I was suffering from near-sightedness and seeing everything double. I had an operation performed by Dr. Agnew, which, with the use of glasses, restored my eyesight and corrected the imperfect coördination. If I attempt to recall scenes that I saw while my eyes were out of order, I invariably see them as they appeared during that time, although I may have seen them many times since the operation. For instance, in the case of the minister in the pulpit at home, I see two images of him, no matter how much I may try to get rid of one of them. My recollections of the examination hall and of the examiner, upon entrance to college, are affected in the same way, although I have since attended several courses of lectures in that room. When I think of the examiner, his several positions are all very clear, but all double. My recollection of the office in which the operation was performed is also of everything as double, although I saw it only twice before the restoration of my sight, and many times after. The objects which I have seen since the operation are always single when recalled."

But we may also have phantasms of touch, taste, sound, and smell. Only a few persons can recall odors ; one writer asserts, on the other hand, that odors are the most vivid of all his recalled sensations. Touches are the next rarest, then sound, then color, while form is most frequently recalled. Of twenty-five writers, all say they can recall form in some degree, and two thirds of these recall form more distinctly than anything else that comes to the senses. Colors, according to this series of replies, can be fairly recalled by about two persons out of three, but not so vividly as forms. With only one fourth the number was the recalling of form and color equal ; with one tenth was the recalling of form, color, and sounds equal. Those who recalled sounds could in few instances recall colors readily, and in many cases there was a vivid recollection of color with a dim idea of form, or *vice versa.* Nineteen could recall form best, eleven could recall colors best, or as well as forms, nine for sounds, three for touches, and two for odors. These proportions probably indicate but roughly those which would be obtained from a larger number of persons. Among individuals they partly attest the relative inborn acuteness of the various senses, as well as individual preferences for certain qualities of objects ; objects of distaste are naturally suppressed from our imagery as far as we can control it ; throughout all is the principle so well brought out by Mr. Galton that our powers of reviving the impressions of different senses are very uneven.

We may likewise have phantasms of purely psychical or mental states, such as joy, fear, hope, reasoning, resolution ; but these have not been so carefully observed, though they are, if possible, of more importance.

SECTION III.

IDEAS SINGULAR AND CONCRETE.

The word " idea " is used very loosely and ambiguously. But it may have a definite meaning. Literally signifying image, it may stand for all those operations in which there is a reproduction of past experiences. When there is an object before me, say a mountain, and I look upon it, I would not say that I have an idea of it, but that I know it. In like manner, when I am conscious of myself in a particular state, say in pain, it is not an adequate expression of the fact to say that I have an idea of the pain ; we have a conscious knowledge of it. But when the mountain and the painful affection are recalled wo may then say that we have an idea of them. That which is brought up by the phantasy may always bo called an idea. So far as it is thus raised it is always like the original perceptions of sense and con- sciousness, singular and concrete, and these may be called phantasms. Out of the singular and concrete cognitions there may be formed general and abstract no- tions, and these may bo called conceptions or concepts. (See *infra*, under Comparison.) Both of these may be called "ideas," according to the usage of the English tongue. `

In an earlier part of this work I have critically exam- ined "the ideal theory." In sense-perception the object is presented and is known directly. When we look at a tree I would not say with Locke that we have an idea of it, but that we have a knowledge of it. But when the tree is not present and we recall it, then it is proper to say that we have an idea of it. We thus see what is the proper order of our mental operations, not first the image and then the substance, but first the substance and then the image. In this way everything is put in its proper place. There are metaphysicians who reverse this order, and put that which is first last, and that which is last first, and thus derange everything, make it impossible to distinguish philosophically between the ideas and the realities, and give to things a shadowy existence. Wo avoid this by making ideas the reflection of things.

CHAPTER III.

To the superficial observer it might seem as if these ever-changing thoughts and feelings of ours follow each other at random. In certain of our moods they leap from topic to topic, certainly with extraordinary rapidity, and seemingly without any order or connection. We would direct them exclusively to some all-important matter, and suddenly they are among objects widely removed and altogether irrelevant. In the midst of business they set off in pursuit of pleasure: when we would compose our minds for devotion, we find, before we are aware, that they are carrying us wandering over the mountains of vanity; while it will sometimes happen that, in our moments of frivolity and folly, the most solemn thoughts will present themselves to sober or to awe us. Our experience thus seems, at least at first sight, to show that our ideas flit, at their pleasure, from gay to grave and from grave to gay; from home to the ends of the earth, and from the ends of the earth back to home; from fear to hope, and from elevation down to flatness; from earth to heaven, and, alas, from heaven to earth. But while this may be our first impression, it will be found, if we inquire more carefully, that just as law rules everywhere in the world of matter over even the most unruly agents, — over the boiling waves, the leaping streams, the fickle winds, — so it also reigns,

with all its order and beneficence, in this kingdom of
mind ; and links, often by invisible ties, our thoughts
and emotions one to another.

We find our ideas pursuing a course. (1.) When we
watch and follow them we find them connected one with
another. Some one refers to the great civil war in
America, and immediately its scenes come before us ;
the circumstances which led to it, the existence of slavery, the feelings of the North and of the South, the bat-
tles and their results ; the terrible sufferings, and the
mistakes committed, the conduct of the statesmen and
the generals, the part taken by Great Britain and
France ; the sentiments of these countries about America, the effect which this had on America, the issue of
the war and the condition in which it left the United
States. Our thoughts have gone over a considerably
wide course, over a number of years, and two wide con-
tinents, but they have not taken a violent leap ; they
have trod the whole way step by step.

(2.) We can often trace them backward, when we find
the same consecutiveness. Often, indeed, we may not be
able to discover all the links. as some of them may be
forgotten in the rapidity of their occurrence. Ordinary
conversation often seems very desultory, yet we can at
times discover the thread on which are strung topics the
most remote and discordant. Thus Hobbes of Malmes-
bury tells of his being in a company in which the conversation turned on the civil wars in the times of the
Commonwealth, when a person asked abruptly, "What
is the value of a Roman denarius ? " "On a little reflection," says Hobbes, "I was able to trace the train of
thought which suggested the question, for the original
subject of discourse naturally introduced the history of
the king and the treachery of those who surrendered his

person to his enemies; this again introduced the history of Judas Iscariot and the sum of money which he received for his reward." I remember trying to make a company merry by the narrative of a fishing excursion which had been distinguished by some laughable misfortunes, — of boastings ending in humiliations, and of duckings without drowning, — when, to my surprise, a lady burst into tears: it turned out that she had lost a dear boy, who had fallen into a deep pool when fishing. In such cases we can detect the train of thought. In others we may not be able to follow the path, as no traces have been left behind in the memory; yet even in such we are certain that there has been a continuous course, just as we are sure that the bullet, though we have not seen it, has passed through the whole intermediate space between the rifle and the target; and that the lightning, which cometh out of the east and shineth even unto the west, has passed through every point between.

> " Who shall say,
> Whence are those thoughts, and whither tends their **way;**
> The sudden images of vanished things
> That o'er the spirit flash, we know not why.
> Tones from some broken harp's deserted strings —
> Warm sunset hues of summers long gone by —
> A rippling wave — the dashing of an oar —
> A flower-scent floating past our parent's door —
> A word — scarce noted in its hour perchance,
> Yet back returning with a plaintive tone —
> A smile — a sunny or a mournful glance
> Full of sweet meanings, now from this world flown;
> Are not these mysteries, when to life they start,
> And press vain tears in gushes from the heart ? "

I am to endeavor to say whence are these thoughts. In doing so I find it expedient, first, to announce and illustrate the laws which are obvious and which are gen-

erally acknowledged, and then to discuss some more subtle and disputed points.   The laws of association are of
two sorts, Primary and Secondary.

SECTION I.

PRIMARY LAWS.

These regulate the succession of all our spontaneous
ideas; not, however, of all our mental states, some of
which, such as our sensations and perceptions, are called
up by external circumstances.   The laws may be arranged under two heads, Contiguity and Correlation.

I. CONTIGUITY.   *When two or more ideas have been in
the mind together, on one coming up it is apt to be followed by the other or others.*   The law takes two forms,
the one that of Succession, when the ideas have followed
each other; the other that of Coexistence, when they
have been together.

(1.) THE LAW OF SUCCESSION.   *When two ideas
have immediately succeeded each other, on one of them
coming up there is a tendency in the other to follow.*
This is the Law of Repetition.   The same follows the
same.   Our thoughts have gone once, twice, or several
times in a train, — A, B, C, D, E; one of them, A, is
started, and off goes the mind after B, C, D, E.

> " John Gilpin was a citizen,
> Of credit and renown."

The child goes over this once, twice, thrice, till the words
have been associated according to the law of repetition;
and now you have only to start " John Gilpin," and
away he slides — as on an icy track which he has made on
the snow, " was a citizen of credit and renown."   Thus
it is, that things having been associated once, twice, or
often in our minds, the one is apt to recall the other.   It

is thus we have joined China and tea; Japan and lack-
ering; Cornwall and tin; Manchester and cotton; Belfast
and linen; Switzerland with high mountains and gla-
ciers; the Highlands of Scotland with hills, brown heath,
rugged rocks, and leaping streams; Ireland with green
grass and foliage; ancient Athens with literature; ancient Rome with conquest. Thus it is, that having often
seen them together, the black coat becomes associated
with the clergyman, the red coat with the soldier in
England, and the blue coat with the soldier in America.
Thus it is that the sign becomes associated with the
thing signified; the rose with England, the shamrock
with Ireland, and the thistle with Scotland. Thus it is,
and as far more important, that when we have become
familiar with the meaning of a word, it at once, and
without an effort, calls up the signification; and in an
hour we comprehend all that is in the lecture with, perhaps, its five thousand words. Thus it is that places
become associated with what has been experienced at
them. We see this law at work even among the lower
animals. If a horse has had a fright at a particular
place it will begin to tremble as it comes to the locality.
The widow, whose husband was killed at a particular
turn of the road, cannot pass it without being overwhelmed with grief. The mother ever remembers her
boy, now, perhaps, grown into a man, as she passes
the place where she parted with him, as he set out to
face the hard struggle of life in some distant city or
foreign land. Thus it is that certain localities suggest
great historical events. Marathon and Bannockburn
and Waterloo call up nations delivered from tyranny;
and Bethlehem and Nazareth and Jerusalem call up
freedom achieved by a mighty deliverer for the sin-
enslaved race of mankind.

8

Taking this key with us we can often explain certain peculiarities of character which may seem very odd. The child screams when he hears of his being about to receive a visit from the surgeon, who had to perform a painful operation on him. That boy will not taste the jelly piece offered him, because the jelly is associated with the nausea of the drug which was administered in it. An excellent lady of my acquaintance was nearly killed by a bullock when a child, and ever since she runs from the most harmless cow as if it were a lion. Thus it is that certain persons have been made to acquire a horrid shrinking from certain objects, such as mice or rats, as frogs or toads, as cats or dogs, or from darkness, which is associated with ghosts. Beginning with these simpler instances, we can now explain more complex and recondite cases ; as how people become prejudiced against certain persons ; these persons have inflicted on them some real, or quite as possible, some imaginary injury ; or against certain scenes, because there they have suffered a humiliation. I know a man who supposes that I kept him out of an office which he very much coveted, and ever since, when he is in danger of meeting me, he sets off the nearest by-way that may enable him to escape. Those who have injured any one in his property or good name are apt ever after to shrink from his company ; for his presence reminds them of their sin, which they would rather keep out of sight. I knew a young man who made a fool of himself, and was laughed at, the first night he entered a debating club, and never after could he be made to face such a meeting, which he always looked upon as an array of bristling spears — the tongue of every member being ready to enter into his heart. Young persons are to be on their guard against falling under the power of such unreasonable or sinful associations. When we are in danger of being subjected to them, we should hasten to deliver ourselves from the thraldom by connecting the objects with other and more pleasing remembrances. I know a boy who in early life got a fright at dogs, and it was only by his being led to mingle for long with very gentle animals, that he was cured of his terror; that is, dogs now became associated in his mind with harmlessness and playfulness. I am acquainted with a physician, who, feeling how injurious it is to children to be in terror of their doctor, contrives to amuse his juvenile patients till he becomes a favorite. It is thus we should endeavor to overcome our antipathies towards all of whom we are jealous; let us think of them under the more favorable aspects of their character, or, if we cannot but know and abhor their bad qualities, let us, at least, ever remember that we ourselves are also sinners. It is thus we should contend against every sinful prejudice; against every prejudice, indeed, except the prejudice against sin, which we should certainly ever associate with loathing and detestation.

But it is of more moment to remark, that it is this law that mainly gives its strength to HABIT. Let us glance for a little at habit and its power for good and for evil. Habit, as every one knows, is characterized by two marked features.

(a.) *There is a tendency to repeat the acts which have often been done.* Certain mental states, ideas, feelings, and resolutions have followed each other in a certain order, once, twice, ten times, or a hundred times; and now, on any one of these coming up, the others will incline to follow — quite as naturally as the stone falls to the ground if unsupported, or as water, bursting from its fountain, will run in the channel formed for it. You

wonder at the drunkard become so infatuated, but the
grieving, the downcast mother, or the disheartened wife,
can tell you of a time — and a sigh heaves her bosom as
she speaks of it — when 'the now outcast and degraded
one was loved and respected, and returned with regular-
ity to quiet and domestic peace in the bosom of the family. But, alas, he would not believe the warnings of a
parent; he did not attend to the meek unobtrusive rec-
ommendations of a wife or sister; he despised the com-
mands of the living God; and, seeking for happiness
where it has never been found, he spurned at those who
told him that the habit was fixing its roots, till now he
has become the scorn and jest of the thoughtless, and
the object of pity to the wise and good : talking of his
kindness of heart while his friends and family are pining
in poverty ; boasting to his companions, in the midst of
his brutal mirth, of his strength of mind, and yet unable
to resist the least temptation. What we see in so marked
a manner in drunkenness has equal place, though it may
not be so striking, in the formation of every other habit ;
as of indolence, which shrinks from every exertion ; and
of avarice and worldly-mindedness, which keep us ever
toiling among the clay of this earth ; and licentiousness,
which wades through filth till it sinks hopelessly into
the mire of pollution : the man is driven on as by a
terrible wind behind moving to fill up a vacuum; as by a
tide with its wave upon wave pursuing each other, under
an attracting power which will not let go its grasp. In
all cases we see how difficult it is for those who have
·been accustomed to do evil to learn to do well ; at times
almost as impossible as for a man, who has thrown himself from a pinnacle, to rise up when he is half way
down ; or for one who has committed himself to the
stream above Niagara to stop when he is at the very
brink.

And let no man try to excuse his criminality on the ground that the acts are now beyond his will. He should resist the wave till it has expended itself: he should seek a more favorable wind to drive him along. He is even now to blame for not resisting the evil and not seeking divine aid to help him out of the pit; and he is chiefly and above all to blame for the habit, which is his formation throughout. For it was by repeated acts — by repeated voluntary acts — that the man wore the ruts and deepened the ruts, out of which it is now so difficult to move him. It was the glass of whiskey or brandy from day to day, the intoxicating drinks from week to week, at the dinner or evening party, — it was this that formed the addictedness to intemperance. In these processes there was criminality at every step; and all that ensues — this slavery and these chains — is a judicial infliction for the evil that has been done : the punishment here, as in hell, adding to the greatness and virulence of the wickedness. In most cases, indeed, the man did not see the consequences, but it is because he shut his eyes to them. He would do the deed only this one time, and then he would stop. But the temptation which swayed him the first time presents itself anew, and once more is yielded to. Having crossed the line which separates vice from virtue, he thinks that a few more transgressions may not much aggravate the offence; he therefore goes a little farther, still cherishing the idea that he may return at any time. At length some rash deed of excess, or unexpected exposure, shows him that it is time to draw back; and then it is that he feels how difficult the retreat. It was easy to slide into the net, but what obstacles catch him as he would draw back. His past motion has created a momentum which impels him farther, and ever on towards the gulf. " Be not deceived, God is not mocked; for whatsoever a man soweth that shall he also reap." He has sown to the flesh, and of the flesh he now reaps corruption. He has sown to the wind, and the whirlwind rises to toss him along as by an irresistible power. He has set the stone a-rolling, and he has to answer for the injury it may do as it descends. He has loosed the wagon, and let it go down the inclined plane, and he is responsible for all the havoc it may work as it dashes on with ever accelerated speed. There are affecting cases, in which the man is conscious of his misery as he sinks — like a traveller lost in the Alps — down the snowy descent into the awful gulf. Take the following confession of a man of genius, a poet and a philosopher, at the time when he had become the slave of opium, taken in the first instance to relieve a bodily disease. "Conceive," says Coleridge, "a poor miserable wretch, who for many years had been attempting to beat off pain by a constant recurrence to the vice which reproduces it. Conceive a spirit in hell tracing out for others the road to that heaven from which his vices exclude him. In short, conceive whatever is most wretched, helpless, and hopeless, and you will form as tolerable a notion of my state as it is possible for a good man to have. I used to think the text in James, that he who offended in one point offends in all, very harsh; but now I feel the tremendous, the awful truth of it. For the one sin of opium, what crimes have I not made myself guilty of. Ingratitude to my Maker and to my benefactors, and unnatural cruelty to my poor children; nay, too often actual falsehood. After my death I earnestly entreat that a full and unqualified narration of my wretchedness, and of its guilty cause, may be made public, that at least some little good may be effected by the direful example."

(b.) *Habit gives a facility in doing acts which have often been performed.* This peculiarity is derived from that just considered. It is the tendency that gives the facility — the acquired momentum that gives the velocity. At first the work could be done only by an effort — only by a special act of the will setting itself to devise means and avoid obstacles. Now, the process once begun goes on of itself. As a consequence, that which may at first have been irksome, because laborious, now becomes pleasant, because easy, — and now natural, that is, according to a natural law.

Under the other aspect of habit, we were led to view its evil results. Now we are rather invited to contemplate its beneficent effects ; and surely the law of habit,· like every other part of our constitution, was appointed for good by our Maker. True, it is found that when we abuse this law it has within itself, and evidently provided for this end, the means of inflicting a terrible judicial punishment. But certainly the law is good to them that use it lawfully. We have forgotten a great deal of our childish experiences, yet we remember so much, and we see enough to convince us that that little boy has his trials at every stage as he learns to read : — as first he masters the letters, one by one ; then the words, word after word ; and then is able, out of these black strokes, to gather a history, or a science, or a doctrine regarding God and Christ, and the soul, and the world to come. And yet how easy do we now find all this as in a few minutes we read a whole page, with perhaps its 1,500 letters ? I mention this for the encouragement of those who are still carrying on their education. For our efforts to improve our minds should not cease with our childhood. We should be scholars all our days on earth ; and until we shall reach the kingdom of heaven, where I suppose we shall also be scholars sitting at the feet of the Great Teacher. I recommend that every young man should, at every particular time, be ambitiously and resolutely engaged at his leisure hours in mastering some new branch of knowledge, secular or sacred. Let one propose to himself to acquire a new language, say German or French; another to master a science, say chemistry or natural history; a third to become thoroughly familiar with some department of civil history; while others, or the same, would make themselves conversant with Bible history, or of the history of the Church of Christ in the early ages, or of the Reformation struggle, with its instructive lessons and thrilling incidents of suffering and martyrdom; or they would master the system of Christian theology, or the plan and reasoning of the Epistle to the Romans. In prosecuting any one of these tasks they will find difficulties; but let me assure them, for their encouragement, that these will be felt only at the beginning, and will disappear and be forgotten, like the difficulties they had years ago in learning the alphabet. And these difficulties being overcome, they will find their minds braced and strengthened by the very effort made and the victory gained. Of all attainments youthful habits of a useful kind are the most valuable — more valuable than even all the knowledge acquired in forming them. And youth is the special time for acquiring habits; habits of industry and application; habits of manliness and independence; habits of activity; habits of benevolence and self-sacrifice; habits of reading; habits of rigid thought; habits of devotion. I have been uttering a warning against the formation of evil habits; but no one will be able to prevent bad habits in any other way than by cultivating good ones. You will not be able to keep down the weeds

except by preoccupying the soil with good seed. And as
I have said, the very labor undergone in forming good
habits will harden the mind and body for further exertion. There is a fable told somewhere of a Norman
captain who became possessed of the virtues, whether
courage, sagacity, perseverance, or whatever else, of the
persons slain by him in battle. This fable becomes a
fact in the history of every one who has acquired a
good habit. Every difficulty surmounted by him, in a
branch of useful knowledge, clothes him with new
strength, and prepares him the better for new conquests.

(2.) *The law of coexistence. Ideas which have been in
the mind at the same time tend to recall each other.* This
law is so allied to the other, that the two might be expressed at once under the general name of contiguity or
redintegration; that is, thoughts that have been together
in the mind, either contemporaneously or consecutively,
tend to bring up each other. But while the two have
affinities, advantages scientific and practical arise from
illustrating the law of coexistence separately.

A curious question has been started as to how many
things we may have before the mind at one and the
same time. Sir William Hamilton maintained that we,
can have a clear idea, at one time, of six separate objects. It is a matter for experiment. You will find, I
think, that if you place before you, in fact or in imagination, a number of objects, — say persons, or marbles,
or chairs, — you will not be able to see or contemplate
more than four or five of them; the rest will either
look very dim, or, if you think of them, you must do
so consecutively. Suppose, then, that you have a few
objects before you, A, B, C, D, E, then on any one of

these, B, coming before the mind, it may call up some
one or all of the rest. You met for the first time, and
conversed with two persons in one company; you afterwards meet one of them, and the image of the other
stands before you, possibly, with the room and furniture in which you talked with them. Happening once
to meet a Belfast man at Rotterdam, I never see him
without thinking of that city. In my childhood I was
accustomed to hear a flock of geese cackling as I lay in
bed in my father's house, and on the romantic hills in the
neighborhood I ever heard the lapwing, the curlew, and
the grouse, and I cannot hear the cries of these birds
now without having the whole scenes of my younger
years before me. In my early life of study I sat in a
room through which the blue flies buzzed most vigor-
ously, and the buzzing of a big blue fly always makes
me seated in a certain room at a little table, with my
Homer before me. You heard a person tell a tale that
interested you; whenever the tale occurs to you, it
brings up the narrator, and *vice versa;* the two formed
as it were one complete thought, and the one hauls in
the other. You have been accustomed to hear a tune
sung to certain words ; whenever the words are brought
up the tune comes up with them, while the tune is apt to
bring back the words ; rendering it very perilous to attach profane tunes to sacred songs, — the tune of " Where
the sweet waters meet " to a hymn, — as there will always be a risk that the worshipper thinks of the "sweet
waters " instead of the Divine Being. As one separate
object thus recalls another separate object with which it
has been associated in thought at any time, so (what is
very much the same thing) if I have noticed a number
of qualities of one and the same object, any one of these
may become a starting-point for an association. If I have

met a man with a snub nose and a blue neckcloth and extremely witty, so that pun flashed after pun, and repartee succeeded humorous description, I am apt, when I see a snub nose, to think of the blue neckcloth and the jokes which were fired off. I was obliged once to sit two whole hours at dinner beside a lady with a blazing crimson gown, loaded all over with jewelry, but who had an awful incapacity for conversation, and I never see that gaudy color without thinking of the lady and yawning as I do so, as I recall the terribly long two hours I had with her, starting topic after topic, without a response. The very idea of a dinner company here suggests to me that the dullest party I was ever at was one where there was a table groaning with rich food, with ten kinds of wine, and all the delicacies of the season, — that is, with very unseasonable lamb in January; and I ever since get terribly alarmed for a stupid meeting when I see a gross display of eatables and drinkables, having no quality but a vulgar and sinful expensiveness.

This law of coexisting ideas helps greatly to give sweep and variety to our thoughts. Were there no law save that of repetition, our thoughts, like Dumbiedyke's pony, would carry their supposed master always by the same route to the same spot. But by the law of coexistence a number of roads are spread out before us, that to-day we may pass into one, and to-morrow into the other. By the law of repetition our thoughts hang on each other like the links of a long chain; by this other they are connected as in a network, branching off in all directions. The law of repetition would carry us rapidly as by a railway to a particular point, but by the law of coexistence we have the freedom of a man driving or riding, or as more independent still, walking, and who may

take the straight road, but who may also take the winding one, and strike off when he chooses from the high-
ways to the byways; or leave all paths behind as he
follows the windings of the stream, and is enlivened by
its purlings, or muses on its dark pools; or as he lies
down by the fountain and gazes on its perpetual spring-
ing and the patch of green around; or dives into the
deep woods, and listens to the eerie sound of the melancholy wind howling through them as if seeking rest and
complaining that it cannot get it; or as he goes out
into the wayless waste, to enjoy a sense of freedom in
wandering at his free will; or as he boldly marches up
the steep mountain to see the sweep of hills and rocks,
of plains and streams, of scattered houses and crowded
towns, with the smoke curling up from them to show
that there are dwellers within.

But in order to take full advantage of this law we
must have the knowledge of a variety of objects, and
this is to be acquired by observation, by intercourse
with our fellow-men, by reading, by travelling, that thus
we may ever have themes to set off upon in our musings
and reflections. Persons in charge of the deaf and dumb
tell us that they may not be more stupid than others,
but being from "one inlet to knowledge quite shut out,"
their range of thinking is very restricted, till they are
taught the use of signs, whereby to communicate with
their fellow-men. The ideas of the uneducated man,
who has never travelled many miles from his native
place, are apt to be very few and confined, and the topics of conversation between him and his wife very soon
become exhausted — persons of intelligence in this walk
of life are commonly so glad when they can have the
society of one beyond their narrow sphere. " I never
thought the world so big till I went to Belfast," said a

good woman brought up twenty miles away in a solitary house in the Antrim glens. The world does look bigger as we come to know that there are other people in it besides those who live in our own parish; or as we study history and study science, and go back through English history, and Irish history, and Scottish history, and Roman history, and Hebrew history, and back beyond to the geological epochs; or as we go out beyond the range we can see from the hills above our mother's house and begin to conceive of so big a country as the United States, and then take in the whole globe of our world, and the sun's magnitude, and realize in thought that these stars are suns and systems of suns; or as we are trained out of God's book to take in the yet nobler idea of a spiritual God, who fills all time and all space, so that they are no longer empty, but full of life and love.

But just because there is such a wide range, there is greater scope allowed our thoughts to wander into forbidden regions. And what starts they do take! You would think of a solemn doctrine of religion, and you remember how you heard it preached by a minister in a particular church, where was a lady, whose character and whose ribbons you find yourself somewhat eagerly discussing instead of the religious truth. This is a small matter, and may be easily rectified; however, there should be a rectification, otherwise we shall soon lose control over our minds altogether. But the danger lies in systematically allowing ourselves to indulge in improper thoughts, which come thus to coexist and mingle with all our trains, so that every train brings them up along with it to carnalize and degrade the mind. Thus there are some who have cherished thoughts of vanity (" how long shall thy vain thoughts lodge within thee,"

— Jer. iv. 14), and encouraged themselves in the midst
of all their employments to think of their supposed ability, skill, prowess, generosity ; and such feelings being
thus fondled present themselves even when they are not
wished, till the man becomes literally puffed up with vanity, which is sure ever to land him in humiliations before his fellow-men, even as the self-righteous spirit has
all along been displeasing to God. Another has allowed
himself to dwell on the evil qualities, real or supposed,
of his fellow-men, till he becomes habitually envious in
thought, and censorious in speech. A third has rolled
impure thoughts as a sweet morsel under his tongue, till
he has defiled his whole soul, and he becomes the easy
prey of the first temptation. In proportion as such
thoughts as these are cherished and mingled with our
whole life, so will they certainly and frequently come up
of their own accord and unbidden. " Be not deceived,
evil communications corrupt good manners." The wise
father is accustomed to warn his son of the danger arising from evil companions, which are perilous in very
proportion as they are pleasant. But there is another
class of companions who are yet more dangerous, because they have yet closer access to us, and these are evil
ideas and evil feelings. We think we can allow these to
dwell in the soul, and yet be untainted by them. Ah,
it is the mistake of the youth who thinks he can go into
scenes of dissipation and folly, and yet keep himself free
from the vices which conquer others. Once having admitted these visitors to our familiar heart, we will not
be able to banish them when we choose. Having called
up these spirits from the vasty deep, we shall find that
we are not able to allay them when we please ; they
will insist on abiding with us, first to tempt, and then
torment us. " Can a man take fire in his bosom, and

his clothes not be burned? Can one go upon hot coals, and his feet not be burned?"

Under this head the youth needs to be guarded against a more subtle seduction. He must beware of identifying morality and religion with what is mean and gloomy; and again, of associating vice or sin of any kind with what is pleasant, and noble, and generous. This is a snare which the wicked will lay in the path of the inexperienced, whom they would tempt to look on religion as fit only for the dying, or as likely to be accepted only by knaves and simpletons; while they represent irreligion as manly, independent, and opening numerous sources of enjoyment. Those who would allure the thoughtless know well how to set off sin and folly by theatrical accompaniments, by the setting of cut flowers which look pretty at night, but which are faded on the morrow, and by scenery which appears fair only when seen in the glare of artificial lamps, but which we turn from as unbearably paltry and shabby in the light of day.

But it is not enough to guard against this association with evil; we must seek the society of the good. And here is the proper place for mentioning, that aids to the memory proceed very much on the principle we are now expounding. Persons connect something which is apt to be forgotten with another thing which must come before the mind, and which, as it comes up, brings its companion with it. This was the use of those signs upon the hands, and frontlets between the eyes, which were used in the East, and which are referred to in Scripture. Such artificial aids are still used in modern education, and commonly proceed on the laws of succession or coexistence. But by far the most useful sort of aid to memory is that which arises from the judicious

arrangement of time, which secures that as the day of
the week comes round, or the hour of the day presents
itself, it lets off, like a mill-wheel, its allotted work.   It
is proverbial that what may be done at any time is apt
to be done at no time ; and the reason why, because it is
not connected with any specific time, and so is forgotten
or postponed ; whereas, when the exercise has been tied
to a particular hour, the hour brings the recollection of
the duty, and the inclination to perform it.   Let me
suppose that there are two young men, one of whom has
made no distribution of his time, but has left himself at
the mercy of circumstances as they cast up, whereas the
other has allotted one part to devotion, another to business, a third to relaxation, a fourth to reading and the
means of mental improvement ; I venture to say that the
latter, with much more ease and satisfaction, will soon
find himself rewarded by having done a far larger amount
of work, and why? because he has conformed to a law
which God himself has planted in our constitution.

It is not enough to accommodate ourselves to these
practical rules ; we must seek to surround ourselves with
the beautiful and the good.   Some have supposed that all
beauty consists in association of ideas.   This I believe
to be a mistake.   There are objects, there are feelings,
which are lovely in themselves.   Still, much of the feeling of beauty which collects around certain objects
arises from their coming to be associated with peace and
plenty, with life or power, or some other living reality.
It is a fact that uneducated persons, and persons low in
the scale of humanity, have, in general, not much sense
of the beautiful.   " That's a grand mountain! " I said
once to an Irish lad who was driving my car.   " Yes,
sir," said he, " it feeds a hundred cattle."   " What a
lovely bank! " said a romantic young lady to a decent

Scotchwoman. "Ou, ay," was the reply, "it is gran' for bleachin' claes." Yet there is surely a pure source of high enjoyment thrown open to those who are capable of looking on multitudes of objects in an interesting light. It *is* an attainment, when we have so cultivated our associations that wherever we are, and in every season of the revolving year, — whether among the opening buds and blossoms of spring, or among the full-blown flowers of summer, or the fruitful riches of autumn, or the pensiveness of the falling leaf, or even among the wrecks strewn by winter, we everywhere fix on objects which awaken feelings of the beautiful, the picturesque, or the sublime. We have a compassion, apt to be mingled with contempt, for the man who looks on Staffa and the Giant's Causeway merely as a lump of rock, or on the ocean simply as a great pool of brine. We have no great respect for the Yankee who lamented that there was so much water power lost at the Falls of Niagara, or for the Glasgow man who valued Loch Ketterin as a reservoir or big tub, holding so many hundreds of thousands of gallons to supply his city with water.

> "A cowslip by the river's brim,
> A yellow cowslip was to him,
> And it was nothing more."

Let us cherish a yet higher ambition. Let us seek to raise ourselves up by the high and noble company we keep, as we surround ourselves with the society of the pure and the good: by holy doctrine; by stern law; by the "primal duties that shine aloft like stars;" by the memories of deeds of courage and perseverance, and self-sacrifice; by images of purity; by models of excellence; by high ideas of God, who is a spirit; by tender, awful, and yet familiar remembrances of the God Man, as He walked the earth and did his work, and bore mys-

terious sorrow while He scattered offices of kindness;
let us surround ourselves by generous sentiments, by
the mercy that is ever pitiful, by " the charities that
soothe and heal and bless," that, as we walk with them
and they converse with us, they may elevate, and yet
humble and instruct and admonish and cheer and con-
sole us.

II. The Law of Correlation, *according to which,
when we have discovered a relation between things, the
idea of one tends to bring up the others.* Attempts
have been made to resolve this law into the others. I
am not to enter into these subtle discussions here; it is
enough for me that there is such a law simple or complex; it is a fact, that things between which a relation
has been discovered suggest each other.

I have classified below (p. 211) the relations which
the mind can discover, those (1) of Identity, (2) Whole
and Parts, (3) of Resemblance, (4) of Space, (5) of
Time, (6) of Quantity, (7) of Active Property, (8) of
Cause and Effect. When we have discovered any one of
these relations the objects are apt to call up each other.
An object in one position calls up the same object in
other circumstances. A part suggests the whole and the
whole a part. Resembling objects bring up each other
as contiguous objects in space and time and quantity do.
The properties and causal relations of objects are powerful bonds of association.

But instead of enlarging on all these we may illustrate a few of them in a way which will come home to
the experience of every one. Thus like recalls like. It
is the law of similarity or resemblance. I see a portrait
and immediately I think of the original. I see a boy
and I am at once reminded of his father, whose features

he bears. This law brings comparisons and likenesses of every kind before us, and we delight to trace them. There are analogies which commend themselves to the minds of all, and the one object ever suggests the other. Thus we connect sunshine and prosperity, night and adversity, light and truth, darkness and error, mist and confusion, whiteness and innocence, sin and pollution, the dove and meekness, the serpent and wiliness, the lamb and gentleness, the tiger and fierceness, the fox and cunning, the dog and faithfulness, fickleness and fortune, the forest and wandering, high winds and calamities, waves and troubles, heights and hollows with prosperity · and adversity alternately succeeding each other, human life and the running stream, spring and childhood, summer and the bloom of youth, autumn and sober middle age, the fading year and declining life, old age and gray hairs with the snows of winter. Prose uses such comparisons for instruction, and poetry seizes them, and brings them forth for delight. "James Thomson," said Samuel Johnson, " could not see these two candles without making an image out of them." The earlier poets brought out the more obvious, the broader, and more striking comparisons, and these ever come home most powerfully to the hearts of all; but these being now become commonplace, certain more modern poets, such as Keats, Tennyson, and Browning, are led to search for more subtle and recondite analogies, which affect most intensely a select few who have run through all older poetry. Poetry seeks to take advantage of all sorts of correlations, of sound and sense, of measured syllables, of rhyme, of balancings of idea and sentiment, of metaphor, simile, contrast, and comparisons of every kind. Hence it is that poetry is more easily committed to memory than prose ; we have now not only the law of

repetition to aid us, but the law of resemblance or correlation, the one strengthening the other, and the whole
giving impetus to the stream.   But this law is not confined in its influence to poetry; it aids the scientific in-
quirer in every branch of investigation, by often bringing together the things that are like, and which should,
therefore, be put into the same class or group.   The
botanist sees a plant; it suggests a like plant, and the
species and genus to which it belongs.   This law of mind
within thus helps us to discover the laws of nature without us ; and to make us feel that we are surrounded
with objects not constructed arbitrarily, nor distributed
capriciously, but fashioned after an ordained model-form
or type, and capable of being arranged in the most
methodical manner into species and genera, and orders
and kingdoms.   By this corresponding law within, we
are thus made to feel at home amid the varied and complicated works of nature ; and to discover among these
adaptations evidence of a plan and a purpose.

And then resemblance is only one of many correlations which the mind is inclined to discover and to follow in its spontaneous trains.   There are a number of
other lines, on which rails are set for it, and on which it
will run if it is once placed on them.   Thus it is apt to
run on from effect to cause, and cause to effect ; from a
whole to its parts, and a part to its whole ; from means
to end, and end to means.   On seeing an effect, the
mind naturally goes after its cause.   Thus, in history,
we inquire what agencies God employed to effect the
great Reformation in the sixteenth century ; what causes
produced the French Revolution of 1790 ; what influence made the people demand the Reform Bill in England in 1830 ; and we would seek to determine what are
the causes that produce the special forms of immorality

which disgrace our country. And again, when we see
a powerful set of agencies at work, we inquire what will
be their effect; what, for instance, must be the issue
of the abolition of slavery in America? In science,
Newton inquired what draws a stone to the ground, and
found it to be the same gravitation as keeps the moon in
her sphere : and there are persons asking what keeps up
the sun's light, and they think it is a shower of bodies
which run round him, and as they ever come nearer dash
into his atmosphere and strike light and heat. Thus it
is that when we fall in with a complex object we are inclined to resolve it into its parts ; and when we see an
ingenious machine, to find out what is its use. We need
not illustrate this farther. All philosophy and all science are illustrations of relations of some kind, of class
or cause, of parts and whole, or means and end; and
minds of a philosophic or scientific character are exercised in discovering these relations, which are the relations that bind nature together ; and all persons gifted
with high intellect delight to pursue and follow such
relations, as they are unfolded to us through all the
works of the Great Creator. Thus, too, we like to have
essays, sermons, lectures, speeches, not disjointed and
leaping abruptly from one topic to a distant one, but, on
the contrary, having all the parts connected by a relation of some kind, and all the sentences leading grace-
fully the one to the other ; and when this is done our
thoughts follow the train more pleasantly, and we find
that by means of the ties thus supplied we can, with
less difficulty, call when we please all the topics back
into our memory.

It should be observed that as the result of the working of the laws of Coexistence and Correlation, our ideas
are apt to come up in groups ; at times to annoy us by

collecting a troublesome crowd; more frequently to help us by calling in the powers which enable us to accomplish our ends, which we now do as if by instinct.

It is by such high relations that the ideas of minds of the higher sort are knit together, and their thoughts, instead of running like those of commonplace minds in the same track, or bringing together the things that have coexisted before, go after analogies and causes and consequences and analyses and uses, and bring illustrations and proofs and confirmation from remote regions. The memory that proceeds by correlation is much higher in kind than that which follows mere repetition and coexistence. It has often been said that a powerful memory is seldom associated with a strong judgment. This is so far a mistake. In order to ascertain the exact truth we must distinguish between two kinds of memory, — the memory that goes by repetition and coexistence, and the memory that pursues resemblances and causes and other intellectual relations. A memory that excels only in repeating may certainly exist without any high powers of judgment or reason. There have been extraordinary instances recorded of this repeating memory. The story told of a man employed by Frederick of Prussia to repeat a poem of Voltaire, on once hearing it, is rather amusing. Voltaire was to read to the king a poem of considerable length, which he had just composed. After he had finished reading the king remarked dryly, " That poem is stolen ; I have heard it before." " That is impossible," said the poet. Whereupon Frederick said he would prove it, and immediately sent for a man who, to the great confusion of Voltaire, repeated the poem word for word. The person had been placed behind a screen, and from once hearing the poem was able to repeat it correctly. But this memory is after all the child's memory, which goes by repeating the same, or striking off after the topics that have been together in its mind before. It is not the memory of the man intellectually advanced, which will not follow the one straight line, because it has many other lines alluring it; which will not spring up straight like the stalk of grass, or the reed, but goes off ramified like the tree in multiplied branches and branchlets of varied curvature, covered all over with graceful foliage. It is not the memory of the historian, the memory of the poet, the memory of the man of science, all of which proceed by correlation, and bring in facts and images and generalizations from the past and the distant, as well as from the present and the near, from the real and from the ideal, from earth and from heaven; to instruct by their truthfulness, to please by their beauty, to strike by their novelty, or to combine the scattered works of God in a sublime unity, illustrative of the one great creative mind. A memory which thus gathers in from such varied quarters is ever associated with, as indeed it proceeds from, a powerful understanding.

SECTION II.

SECONDARY LAWS.

THE LAW OF PREFERENCE, WHICH IS THE LAW OF NATIVE POWER AND ACTIVE ENERGY. — The laws of which I have been speaking seem to me to be those which regulate the train of thought. But the question arises, How is it that the mind is led to follow one of these rather than another? or why, among a variety of objects which it might follow, does it take one rather than another? I met two persons in a particular company; the next time I fall in with them I remember the one, but not the other.   Why is this?   The reason may be that the one of them had a very brilliant conversation, or he committed some great blunder which exposed him to ridicule, or he had a pair of peculiar gray eyes, or a limp as he walked.   The question now is, Can we generalize these reasons?   The laws of this kind have been called Secondary Laws by Brown, and the Law of Preference by Hamilton.   I have a way of my own of stating them.

There seem to be two grounds on which the mind turns to one associated object rather than another.   The one of these is the ground of Native Power, Taste, and Disposition.   We are so constituted by nature that our feelings go in one way rather than another.   Thus one mind ever tends to repetition, another rather to correlation.   One man delights in poetical images, another in scientific classes or causes.   One intellect is inclined to observe resemblances, another differences and exceptions. I cannot illustrate this; it does not bear so much on the practical object I have in view.

But I must explain and illustrate the Law of Mental Energy.   Those ideas are brought up most readily and frequently on which we have bestowed the greatest amount of mental exertion.   Thus it is when we have once and again, in the past, thought or felt about certain objects, they will be apt once and again to come before us in the future.   Every mind seems to be endowed with a certain amount of power, and according to the power expended on an idea, so is it remembered for a greater length of time, and it comes up more easily and frequently.   I suppose it is because youth has the greatest amount of this energy that the memory is strongest at that period of life, while in consequence of fading strength old persons feel a less interest in the objects and events which pass before them, and these in consequence leave little impression on their minds. This exertion may be an energy of *Feeling*, of *Intellect*, or of *Will.*

(1.) *Those ideas that have been attended with deep feeling are called up more frequently and readily.* — I have forgotten many of the events of my childhood, but there are some I can never forget, they were accompanied with such deep emotion. I cannot, for example, forget that solemn sabbath-day in which I saw the corpse of a revered father spread out on the couch on which I had so often played with him. Much that I did and saw in my college days has passed into oblivion; but I cannot banish one scene from my mind. I was passing along the streets when I saw a child literally divided in two by the wheel of a heavily-loaded cart passing over it. I got but a momentary glimpse of the scene, of the mother hurrying frantically past me to embrace a mangled corpse instead of a living child; but there it remains painted on my memory, fresh as if it had happened but an hour ago. I do not remember all that I saw in a pedestrian tour which I took in the highlands of Scotland, but I can never, while I have a memory, forget such scenes as Loch Lomond and the Trossachs. I cannot recall all that I saw in Germany, but I can bring up at pleasure the "Unter den Linden" of Berlin and the scenes hallowed by the deeds of Luther. If Italy is named, Venice, and the plains of Lombardy, and the Cathedral of Milan, and the Lake of Como come up lively as a picture. I remember much of Switzerland, but I dwell most fondly on the sunrise as seen from the Riffelberg; on the high top of Mont Blanc, flanked with its pointed buttresses; on the huge bulk of the Jungfrau, with its deep gullies; and on the placidity of Lake Lucerne, guarded by its horrid mountains of snow and ice. My recollection of Belgium is somewhat flat, but I do not forget Waterloo. Why do such scenes and events, and a thousand others of the same kind, rise up like mountain-tops in my retrospective memory, while others have sunk out of sight like the valleys between? It is because, to use a geological illustration, they have been heaped up by the fiery heat of deep and fervent feeling, which has elevated them from the usual low level of life to cut and to face the sky.

We see, then, one way of preserving events in the memory. We may let the mean and the trivial pass away into oblivion. But let us preserve those that are worthy by embalming them in warm feeling. You can often determine what a man is wont to feel an interest in by the objects which he remembers. Suppose a number of persons, of different tastes, training, and trades, have traveled over the same country in company, you may guess the objects in which they felt an interest by the nature of the scenes remembered by them most vividly. The farmer has a distinct recollection of the soil, and of the way in which the land is cultivated. The merchant and manufacturer will rather dwell on the symptoms of advancing or declining trade in the towns and villages along the route. The man of scientific culture can tell you what were the plants and animals of the district, and what the structure of the rocks; while he who has a taste for the beauties of nature can never forget those hills and glens and streams of romantic beauty which so kindled his eye as he passed them. The antiquarian delights to describe that ruin covered by the hoar of age; while the man of historic taste will wonder at all the others because they never noticed that plain where a great battle — that decided the doubtful fate of a country — was fought, or that house which was the birthplace of some patriot or poet. And that plain-looking dwelling: it was not observed at all by the mass of the company, and those who noticed it thought it one of the dullest, stupidest places on the whole journey; but to one man it is one of the kindest and most endeared spots on the surface of this wide world, — for that house was his birthplace, associated with his earliest and dearest recollections, recalling the scenes of childhood and the countenances of friends departed from this world, so that the wealthiest cities on the route, and the most gorgeous temples, and the loveliest of the valleys, and the grandest of the rocks and mountains, have not had to him half the interest which this place possesses.

(2.) *Those ideas come up most frequently and readily on which we have bestowed the greatest amount of intellectual energy.* — Children, it is well known, are apt to forget those lessons which they have learned easily, whereas other lessons acquired with greater care cling to the memory. We sometimes see the principle very strikingly illustrated in the case of two boys in the same family, one of whom learns quickly and forgets as rapidly, whereas the other has acquired his task more laboriously, but retains it longer. There were many advantages in the old plan of thorough drilling and disciplining. I have no faith in science made easy, and philosophy in sport. Some one was recommending to Sir Walter Scott a plan of teaching science by cards. " You will easily," he remarked, " teach them to be fond of the cards ; you will have a greater difficulty in giving them thus a taste for the science." I have no faith in those quack teachers who can make you master of penmanship in twelve lessons, or of French in three months, or Latin in a twelvemonth. There is really no royal road to knowledge. The Prince of Wales must learn his mathematics in the same laborious way as the peasant's son. Drilling is a good thing in itself; it is the sole way to make an intellectual soldier ; only the exercise, if not less laborious than it used to be in old times, should be made as interesting and pleasant as possible. There is no help for it; the man who would get to the top must climb the mountain, but there may be some rests by the way, and he may get some pleasant views as he ascends. Young men should not grudge the labor bestowed on a branch of study ; it is one of the conditions of their being able to retain what they have got, for it is as true of knowledge as of money, that what is rapidly earned is often as rapidly spent, whereas what is laid up with care and industry is commonly more sedulously preserved. Young men cannot acquire a more valuable habit than that of giving their intellect thoroughly to their business, their reading, their religion. There are many persons who, from neglecting so to discipline their intellects, seem to have lost all power of actively exercising them, and any knowledge they have acquired has passed through their mind, like the familiar striking of a clock, or like the walk from their house to their place of business. Their very reading, which is chiefly in novels and romances, is a sort of idleness, which gives no robustness to the frame ; is often, indeed, a sort of intoxication, which exhilarates without strengthening, and ends in *ennui* and disgust. Those books are the best, not which think for you, but which make you think. If you have not thought over what you have attained, all your acquirements are like the articles in a lumber-room ; some of them may be valuable enough, but they are not at your command. It is when you have thought a book and its subject all over again that it becomes your own, and its thoughts will not only abide in your memory, but be, as it were, incorporated into your intellectual being, to abide with you for strength and for useful application.

(3.) *Ideas come up more readily and frequently when they are associated with an act of the will,* more especially when we exercise an act of *Attention* regarding them. It is mercifully, as I think, provided that much of what we have witnessed and experienced passes away speedily from the memory. Were this not the case, our minds would be filled with innumerable trifles. We must hear the timepiece that strikes in the room in which we are, but a minute after we cannot say whether we heard it or not. How little does a commercial traveller re-member, in ordinary circumstances, of his journey from Manchester to Liverpool, or from New York to Phila-delphia, which he has so often taken! It is well that much passes away after this manner. But we must not allow all to vanish in this way. And we have a means of retaining what is valuable : let us exercise our wills regarding it, let us direct our attention towards it. We blame our memories when we forget, but there is often a greater culprit than the memory, and that is the heart; we felt no interest in the subject; the party to blame is the will; we paid no attention to it. I know a man noted for his forgetfulness; but it has been re-marked about him that he only forgets what relates to his neighbor's interests and feelings; he never forgets what relates to his own. We must all know children who are forgetful enough when we cannot get them in-terested, but who never forget what is taught them when we can allure them to attend. We complain that we are apt to forget the books we read, the sermons or

lectures we hear. Possibly it may not be of great importance to remember all that we have read or heard:
but if it is worth recollection, there is a means of fixing
it. Our voluntary determinations have a sort of anti-
septic power to preserve what they are applied to. Let
us, as we read the book, voluntarily recall the topics at
the end of every chapter. Let us go over mentally, after
it is closed, the sermon or lecture we are anxious to retain in the mind ; or, better still, let us take notes of
the topics of importance in the book or lecture, and we
have got the patent process which will fix forever the
colors that might otherwise fade.

The question has been asked, How is it that when we
form a purpose to do an act at a certain time, we recollect, in the multitude of our thoughts within us, to perform it? The answer is that we are able to do so only
when the resolution has been sufficiently earnest. If
formed in a careless way it may never be executed. We
have heard of the young gentleman who forgot the
appointment he had made to meet a young lady at a
particular hour. Whereupon she cast him off, very
properly, for if his love had been deep he would not
have been so oblivious.

We live in an age in which men know well how to
use all sorts of material power, how to use water power
and steam power and electric power; and they guide the
steam, and condense the vapor, and place wires to conduct the unseen agency, and all that they may set up
incalculable machinery wherewithal to produce nutri-
ment and covering, for utility and for ornament. But
God has given to every one a lease of a far more important power, that we may guide it into the proper channels, and get it up at the proper times, and direct it along
the proper lines, and all that we may awaken genuine

feeling, and gather swift knowledge from afar, and go on to useful and benevolent action. But youth, I remark, is the season in which this power is the quickest and the strongest and the most easily directed. In after life we shall be apt to find it already directed in channels from which it cannot easily be moved; and (to change the image) the endeavors you make to get up life will be like the attempts of the birds in October to raise a song: a cheerful note it is in its way, and we do enjoy it at such a season, but it is not like the full chorus of the wood in spring, — and such is the activity of youth, when it is wisely directed, and all turned into a song of praise to the Great Creator.

We have thus shown that law reigns in mind as it does in matter. When we know what the laws of matter are, we can take advantage of them, and apply them to useful purposes in the arts. When we know what the laws of mind are, we can apply them in the education of the mind.

But before closing I must guard against an impression which may be left, when it is proven that the succession of our ideas is governed by laws which operate independently of us. It may be concluded that we have, and that we can have, no control over our thoughts and feelings, which move, and must ever move, like the winds of heaven. I have been laboring to give the very opposite lesson. It is because the succession of our mental state is under law that we can command our minds and bring them into subjection. We certainly see many who seem to have as little control over their own minds as they have over the minds of others; they are the slaves of the thought, the impulse, the feeling, the suspicion, the passion, that happens to come up at the time or be uppermost. But we have a will, and a free will, given us by God for this purpose, that we may rule our thoughts; and this we can do most effectually when we know what the laws are which our thoughts obey in their order and succession. We cannot, indeed, will into our minds any absent thought; for, as has often been shown, to will it is already to have the thought. If I have forgotten the name of the capital of Japan, I cannot command it to appear. But if I remember any object with which it is associated, I can, by an act of the will, detain this, and think of it till what I want comes. I can think of Japan and of the Japanese I have seen, till Tokio comes up by the law of association. It is for this purpose I have been at such pains to expound the laws of association, that as knowing them we may employ and apply them for the proper ordering of all our knowledge, for the formation of good habits, and generally to obtain a thorough command over our minds, — a command which we find to be more glorious than that of the general when he has horse, foot, and artillery, all so trained and disposed that they move like the limbs of his body at his will. "He that ruleth his spirit is better than he that taketh a city." By thus systematically disciplining our minds, we shall find that we have a greater control over our thoughts than we at first imagined. We shall find that as we habitually repel them, the things that are vain and evil disappear, while the things that are great and good, as we cherish them, remain with us, to talk with us, to instruct us, to elevate us. He who has a mind so stocked and trained is like the traveller who carries his provisions with him; he is in some measure independent of the ordinary accidents of life and the circumstances in which he may be placed, for he can feed, wherever he goes, on the stores he has laid up.

SECTION III.

PHYSIOLOGICAL PROCESSES INVOLVED IN ASSOCIATION.

The question is put whether I have explained thoroughly the facts of association. I answer that it is doubtful whether this has been done by any one. There is one important element of which no account has been taken. Association of ideas must depend partly on the brain, on the gray cellular matter at the periphery, or on the currents through the brain, or, as I rather think, on both, the nature and disposition of the cells determining the direction of the currents.

There is reason to believe that every action of the mind, intellectual and emotive, leaves an impress on the brain. It may be maintained that a concurrence of brain action is necessary to mental action, particularly to the calling up of the ideas of material objects. Our ideas flow more pleasantly when the brain is in a healthy state. On the other hand, we feel embarrassed in the ordering of our thoughts when oppressed with headache: we labor to call up a series of ideas, and we find that they will not appear. Late at night, after hours of anxious thought, we find that the required train will not move on ; it will start on its journey only after we have been refreshed by sleep. It is clear that the cellular powers or nervous currents must help or hinder the use and flow of ideas.

I believe that every thought and every feeling produces an effect upon the cellular portion of the brain, and leaves an impress upon it. Now, in order to the reproduction of the thought and feeling in memory, it seems to be necessary to have a coöperation of the organ of the brain thus affected, and to have the aid of currents. When the association has not this concurrence it is hindered and restrained. How irksome do we find it to learn the grammar of a new language, or the technicalities of a new science, or, indeed, to penetrate into any unfamiliar subject ; while we find it easy to use the tongue, or follow the science, or to speak on the topic, when lines have as it were been made for us in the brain to carry us on. So, when there is a lesion in a particular part of the brain, we may lose certain of our recollections, say of Greek, or of certain events in our past life. In old age memory is the first faculty that fails, because of decaying or decayed organs. The recollection of names is apt to go first, because, names being commonly arbitrary, there are no mental correla-

10

tions to bring them up, and we have to hunt for connected ideas to call them forth. I have observed when names have correlations, when they are titles or are expressive of the objects, they are as easily brought up as other things.

Physiology has to advance several stages before it can give a full account of the connection of the brain with the use of thought. We should be grateful to it for any light it may throw on the subject. But two important positions are to be defended. *First*, The ideas in the mind are not mere cellular or nervous products. We cannot perceive them by the senses. The microscope has not detected them. We are conscious of them, and our consciousness tells us their nature, which is mental, and not physical. *Secondly*, There are mental laws of association, such as I have just been seeking to enunciate and illustrate, say Contiguity and Correlation. These are undoubtedly the principal laws guiding the flow of our ideas ; the physiological ones being merely subsidiary.

This may be the most appropriate place for noticing the circumstance that as trains of thought, at first conscious and voluntary, are confirmed by frequent repetition, they become more involuntary, and we are scarcely conscious of them. Thus we come to run over the letters of the alphabet and the numbers 1, 2, 3, . . .100, without an effort, with scarcely any feeling, and with no recollection. It is supposed that many of our organic actions were originally voluntary, but are now involuntary and unconscious. I am convinced, however, that much of this action has still a sort of dull consciousness attached to it, and that the dormant will may awake on occasions. This may account for some of the curious phenomena of our compound nature, such as mesmerism, dreaming, and so forth.

Measurements, not always trustworthy, have been made as to the time occupied in reflex action, as when a sound or sight goes up to the brain and is answered by speech. But it has been found more difficult to determine the time occupied by our purely mental acts, say by a succession of ideas in counting. Is there any relation between the normal time of the successive ideas in our mind and that of the beating of the pulse and winking of the eyes ? It is certain that the flow of ideas differs very widely in different states. In fever the rapidity may become very great.

SECTION IV.

THE RAPIDITY OF THOUGHT.

Lord Brougham has given us instances of the rapidity of thought. He was dictating when he fell asleep while his clerk wrote the sentence he had dictated. On awaking he found that an immense number of thoughts had passed through his mind. But we have now more accurate measurements. I have been favored with the following summary by James Mark Baldwin, A. B., Ex-Fellow of Princeton College. (See his excellent translation of Ribot's "German Psychology of To-Day.") The measurement of the duration of mental acts was begun by Donders about 1861. Before him, it was generally admitted that psychic processes must be construed in time, and the question of the rapidity of thought was discussed from a standpoint of consciousness. We think sometimes faster, sometimes more slowly. But this subjective estimation of time was necessarily vague, inasmuch as it was impossible to eliminate the physical and emotional influences which alter the flow of our ideas. Since the discoveries of Helmholtz and others, as to the velocity of nerve transmission, it has become possible to arrive at a determinate expression for the time necessary to some of the simpler processes.

(1.) Beginning with sense-perception, the simplest intellectual act, the case is briefly this : Let the skin of a man in normal condition be pricked, and let the subject speak as soon as the pain is felt. The period which elapses is called the simple reaction time, and is found to vary with the different senses from one eighth to one fifth of a second.

Upon consideration it is readily seen that this period may be divided into three parts; first, sensor transmission to the brain; second, the mental process of perception and volition ; and third, motor-transmission to the organs of speech. Now since the velocity in both the motor and sensor nerves is known, we reach by subtraction the time of the mental act. Instruments are used by means of which differences to the ten-thousandth of a second are noted. Avoiding figures, which are still somewhat in dispute, we may give two general principles.

(a.) *The simplest mental act occupies an appreciable period of time.*

(b.) *The purely physiological time is less than half of the entire reaction.*

(2.) Passing from simple perception to the reproduction of ideas as memory pictures, it is concluded from experiments conducted upon similar methods, that

(a.) *The time of the reproduction of a state of consciousness is longer than the time of its production.*

(b.) *The time of reproduction depends upon the degree of energy exerted* (1) *in the original perception,* (2) *in the reproduction.*

(3.) A third operation, upon which many experiments have been made and definite results obtained, is that of *discernment* or discrimination. Two colored lights are shown indiscriminately and the subject is to react only when he sees the color agreed upon beforehand. This involves first a comparison and second a judgment. By an easy process the purely physiological time is eliminated and the duration of the mental act is found to be one twentieth of a second (Kries) to one tenth of a second (Wundt).

(4.) Experiment has rendered service also in defining and confirming the laws of association. The time of a simple association is determined, that is, three fourths to four fifths of a second.

(5.) A fifth class of experiments relates to the logical judgment of subordination (from species to genus). It is found that the time is longest when the subject is abstract and the predicate a more general notion ; shortest when the subject is concrete, and the predicate a less general notion. The average of a great number of experiments gives the time about one second.

It should be said that these results are true only in an average sense and under normal conditions. During the last five years great activity has been shown in the study of abnormal and artificial states, but the difficulties are very great, and the present condition of the science does not warrant a positive statement of results.

It may be added, however, that in every case the general utterances of the inner sense are directly confirmed, and the ultimateness of consciousness as the psychological point of departure is in so far vindicated.[1]

SECTION V.

DISCUSSIONS AS TO THE LAWS OF ASSOCIATION.

I have illustrated the subject in the loose way in which it is commonly presented. But difficult and disputed points have arisen. All

[1] Books of reference on this subject are : Wundt, *Physiologische Psychologie*, ii. cap. 16 ; Ribot, *German Psychology of To-Day*, Eng. trans., chap vii. ; Buccola, *La legge del tempo*, etc.

are agreed that Contiguity is a law of·Association. Some reckon it the sole law, and argue that by some little subtlety all other laws, such as that of Resemblance, can be reduced to it. It will also be generally allowed that there is a law of Correlation to the effect that, having discovered a relation between objects, when the one comes up the other is apt to follow. But it is evident that this may be merely an exemplification of the Law of Contiguity, for the objects have been together in the mind.

It is clear, however, that the correlation discovered greatly strengthens the association. We remember a discourse much more readily when the thoughts were connected. It is for that reason we can commit to memory a piece of poetry more easily than one of the same length in prose; in the former case we have not only the con-tiguities but the congruities to carry us on ; we have the correlations of sound and sense. There is an amusing story told of a minister who, on finding a boy at the helm guiding a vessel, inquired of him if he could box the compass, which he did. He then asked him to do the same backward, which he also did. The boy then asked the minis-ter to say the Lord's prayer, to which the clergyman complacently assented. The boy then insisted on his saying the Lord's prayer backward, which he declined to undertake. The boy was able to interpret the instrument backward because in the compass he had correlations, whereas the other had none in the Lord's prayer. Sci-entific truths are more easily called up than scattered, disconnected ones, because they have been placed under laws of correlation or connection.

But the question arises, Do correlated things suggest each other before the correlation has been discovered ? On entering a room we see a portrait on the wall, and we immediately think of the original, whom we have often met. Had we ever seen the original and the painting together the idea would have been called up by the Law of Contiguity. But we never heard that there was a portrait of the person, and yet his figure casts up. Apparently it does so by the law that " like recalls like." *Prima facie*, the Law of Resemblance seems a simple and original one, and has commonly been so regarded.

It should be noticed that, in order to correlative association, the two objects must both have been previously in the mind ; the por-trait is before us and we are acquainted with the figure and expres-sion of the original. In seeking to penetrate deeper into the nature of the suggestion of resemblance, it is to be borne in mind that objects are known by their qualities, and that all actual objects are complex or concrete, that is, have several qualities, and they are so remembered by us. Let the letters of the alphabet denote the objects related. Let us denote the portrait with its qualities as a, b, c, d, in which a, b are the figure and expression and c, d, the canvas, frame, etc. This portrait recalls the person a', b', x, y, etc., when a', b' are the features and expression, and x, y, etc., the man's walk and gestures. Now it may be argued that a, b of the portrait call up a', b' of the person, while the others, x, y, come up according to the Law of Contiguity. If this view be correct it is the same that calls up the same, the second same calling up by contiguity the objects associated with it.

It may be more difficult to explain in this way other correlated associations. But let us try some of them. In doing so we may find that every relation has a ground, and that they are the same qualities in each object forming the correlation that constitute the principle of the association. It should be observed, however, that it is not by the affinity of abstract qualities that the association takes place, but simply by the objects possessing the same qualities; by the portrait and the original both possessing the same qualities, figure, and expression, in this respect being alike.

We can account in the same way for Contrast, being, as Aristotle asserted, a law of association. Contrast, as a relation, comes under general correlation of Resemblance and Difference. In all Contrast there is implied some sameness ; there is no contrast of things entirely different, and the implied sameness a, b, in both binds the objects together in our minds.

This seems to be the law of correlative association. The same suggests the same, which by contiguity brings in correlative objects, and the relations are perceived by the mind. Those acquainted with the lectures of Thomas Brown, of Edinburgh, will remember that he has two kinds of suggestion, Simple and Relative, — Simple being much the same as I have been describing in Chap. III. Sect. 1. But Relative suggestion embraces two powers : the one association proper, and the other the discovery of relations. These I think should be carefully separated. The latter is really the power of discovering relations or comparison. But while they are different, they may combine in the way I have been endeavoring to describe, and the process may be called Relative Suggestion.

Let us view the mind acting under this power. When objects

present themselves to us by sense or by image, they do so by means of their qualities.   Our anxiety is to know what the object is, and how it stands related to other objects.   Whence has it come? How does it act?   As we keep the object or idea before us, one associated quality after another presents itself, all, it may be, in an immeasurably short time, — we say, "quick as thought."   As they do so, the mind perceives by its power of comparison various relations, and as the result we find what the object is, what its nature and its use.   That cry is the same as I heard in my boyhood, in the mountain region I used to visit; it is the screech of an eagle.   That sound is of a bell inviting me to the house of prayer.   That picture has the features and expression of a friend I knew well, and is his portrait.   The wound of that person lying on the ground is the same as I have seen inflicted by a gun-shot, and I fear the person has been murdered.   A boy is going along a road with a satchel of books; this suggests a school, and we decide that the boy is going to school.   We may have noticed that it is only after allowing the object as we think of it to suggest one quality after another, that we touch the chord which discloses to us what we are in search of, — the nature and use of the object.   Thus closely are the associations of correlation and the discovery of relations connected together and mutually aiding each other.   But the farther discussion and illustration of this subject and its application to cause and effect, to identity, and other relations, may be expediently deferred till we come to discover the nature of Comparison and Relation under Book III., The Comparative Powers.

It should be noticed that a great many of our associations are carried on by means of words.   These words are primarily associated with thoughts and things by the Law of Contiguity.   But each of them is associated with other things which are brought into relation with each other in our minds.   The orator is enabled to carry on his speech without a break, the thoughts and words mutually suggesting each other.   We may often notice a very beautiful play of association in the conversation carried on by a company of intelligent and witty people, each starting and pursuing suggestions with their numberless correlations.

I am here giving as important a place to Association as those who, following David Hume, have accounted by it for our conviction as to cause and effect and the deeper principles of the mind.   A large body of profound philosophers maintain that there are necessary principles in the mind, such as that requiring us to believe that an effect must have a cause. The school to which I have referred, and which I may call the School of Hume, or the Empirical School, explains this by invariable association: the cause and effect having ever been together, we cannot think of the one without also thinking of the other. Now, it is undoubtedly true that when things have been invariably together in the mind in the past, the one will recall the other. But this is a very different kind of necessity of conviction from that which is attached to fundamental truth. It can be shown that this last proceeds from self-evidence, is seen to be in the nature of the thing perceived, and is perceived by the reason. We perceive that it is in the very nature of the cause to produce its effect; for example, of fire to burn. The Law of Contiguity may produce invariable associations and make one thing to come up after another in the mind, but cannot produce necessary convictions or judgments pronounced on a discovery of relations in the nature of the things.[1] It is now acknowledged that mere Contiguity cannot give us *a priori* truth, and we have a new theory that this is gendered by heredity, of which all I have to remark here is, that it may give us tendencies of thinking, but certainly not the decisions of reason. But while Association (and heredity) cannot do this, it may aid our comparative or judging powers by bringing before the mind the ideas on which they pronounce a judgment. I have shown elsewhere ("Logic," pp. 166, 167) that Association brings together, more especially by the Laws of Correlation, the notions, major, minor, and middle, which are compared.

[1] See an admirable history of the discussions in regard to the association of ideas, and a sifting examination of the attempt to account by this for our necessary principles, in *La Psychologie de l'Association*, par Louis Ferri.

CHAPTER IV.

THE RECOGNITIVE POWER.

SECTION L

ITS NATURE.

It is the power by which we recognize an object as having been before us in time past. Let us bring out by analysis what is involved in this capacity.

(1.) *We recognize an object.* — In this there is more than a mere image, phantasm, or idea. That object does not come under our notice for the first time ; we recognize it as having been before us at a previous date. That object may have been a material one perceived by the senses, or it may have been a mental state or consciousness, or a judgment passed or a feeling experienced. Quite as frequently it may have been an event occupying more or less time, and with more or fewer details.

(2.) We recognize the object *as having been before us.* — We not only remember the object; we remember it as something which has been under our notice before ; we remember it as having been in our consciousness. These two elements are in the concrete recollection, and we must give a place to both if we would unfold all that is in the mental act. All our recollections are memories of ourselves and of our experiences. This analysis may, on the first hearing, sound as if too subtle. And it is to be acknowledged that in the ordinary exercises of memory from day to day this perception of ourselves is not prom-

inent. The fact is that though it is in all our memories, we are usually so absorbed with the event that it
is scarcely noticed. This is one of those cases in which
an element of a concrete act very much disappears because we are occupied with the other or others. Still
this element is always present. In every act of memory
proper (not necessarily of the phantasm) we know the
object or event as having been previously before us.

3. We recognize the event as having been before us
in *Time Past.* — It does not come before us in an uncer-
tain way as to its occurrence, as to whether it is past,
present, or future. We regard it as past ; we believe it
to have happened in time past. In proof, we appeal to
consciousness, personal and universal. This introduces
us to

SECTION II.

THE FAITH ELEMENT.

In all these reproductive acts we believe in the previous existence and previous knowledge of an event which
may not now be present, but was before us in the past.
Here, then, is a primitive faith, as distinguished from
primitive cognition in which the object is present.

We draw the distinction between faith and sight. It
is a loose and popular one, but it may be made a philosophic one between primitive knowledge and primitive
faith. In the former the object is present and known ;
in the latter it is not present, but is believed in. This is
the distinction between the simple Cognitive Faculties
on the one hand and the Reproductive on the other. In
the one, the object is present and is known as present ;
in the other, the object is not present, but is recognized
as having been present at a previous time ; in short,
is not presented but re-presented. There would be no

propriety in saying of our immediate sense-perceptions and consciousnesses that they are acts of faith, for the objects are before us and known. When I receive a blow from an instrument and suffer pain, it would imply a confusion of thought and an abuse of language to say that I had a belief in the instrument and the pain; we are giving an adequate expression of our experience only when we affirm that we know the objects. But it would be proper in narrating the occurrence afterward to declare that I believe in the existence of such an instrument, and that I suffered from the blow inflicted by it.

We have now come to a belief, in a rudimentary form, in the absent and unseen. This is an essential part of our nature. It is a most important element in our constitution, standing next to our power of primitive cognition, and in some respects higher than, and certainly prior to, our discursive or reasoning capacity. There are some who insist on our proving everything. They forget that as we can prove only by means of premises we must at length come to premises which cannot be proven, and which must be assumed as being either primitive cognitions or primitive faiths.

If it be asked why we believe in the trustworthiness of memory, the answer is that it is a case in which we are not entitled to ask the why. There are cases in which the mind feels itself entitled, nay, required, to ask a reason. If I am required to give credence to a story about Romulus being suckled by the wolf, I demand proof. But I need no mediate evidence to convince me that I am seated on a chair as I write this, or that before writing I had thought over all these subjects. I feel that any proof proffered would not add to the strength of my conviction; would, in fact, be an impertinence. The evidence — if we can call it so, and I think we can so call it — is not mediate, but is in the very cognition or a belief in the thing itself, and is called immediate, not simply because it is in the perception, but is in the thing perceived. I require proof when it is asserted that the dog star is a certain distance from the earth; and when I get it I am satisfied. But I am equally satisfied, without external proof, that I cannot rise from my chair and go to another, without passing through the space between. In all investigation, if we follow it sufficiently far, we come to such primitive rocks. He who would go deeper down is trying to get beneath the foundation. He who would go farther back is trying to mount higher than the beginning. Setting out with these primitive truths we find their accuracy confirmed, but not primarily established, by our experience. We remember the hills and valleys where we were brought up, and on returning after many years we find them corresponding to our recollections.

The faith before us is of a primitive kind, but it is the beginning of those faiths in the past and in the future, in time and in eternity, which mount so high and carry us above and beyond our world and our experience. We should find pleasure as we advance in noticing the origin and nature of these higher beliefs. Meanwhile we are invited to notice how faith comes in. So far as the initial faith is concerned it is a primitive belief in objects primitively known, — it is the atmosphere that compasses the solid earth.

SECTION III.

THE IDEA OF TIME.

We see how this idea arises. Every event remembered is remembered as having happened in time past.

This gives us the idea in the concrete — "an event having happened in past time." We can now, by a process of abstraction, separate the time from the event, and we have the abstract idea of time. As we do so we are sure that the time is quite as much a reality as the event that has occurred in it. I am sure that I was at a particular dinner party, but I am quite as sure that it was at a certain past time. If it is asked, What sort of reality has it? I answer that it has the reality which I am led to believe it to have. It is not known by me as the substances mind and body are, as having potency. But it is known as having being and independence of my observation of it. It is known as a thing in which events occur, and that the time is a reality quite as much as the events occurring in it.

We are now in a position to criticise the opinions as to time which have been entertained by distinguished philosophers. Locke, as we have seen (p. 84), derived all our ideas from sensation and reflection. He evidently saw that he could not get the idea of time from sensation and so drew it from reflection. We reflect, he says, on the succession of events and thus get the idea of time. But, I ask, how can we know that there is a succession except in time, of which therefore we have some knowledge. To know one event as following another is already to have an idea of time. Here, as in so many other cases in which metaphysicians are endeavoring to simplify the operations of the mind, they are simply assuming what they profess to prove or explain.

At this point Locke has, I think, been successfully met by those who maintain that the mind itself, in its exercise upon the materials supplied by the senses and consciousness, is a source of ideas. Leibnitz and Kant showed that the idea of time could not be had from the experience of sense or consciousness. But their theories on this subject are as objectionable as those of Locke. Proceeding on the principle that the ideas of space and time could not be had from sense, Leibnitz made them mere relations between objects and these relations given by the mind. Kant, proceeding on the same principle, represents them as being forms given to the objects by the mind, thus making them entirely subjective. Fichte followed, and argued that if the mind could create space and time, it might also generate the objects discerned in space and time, and this led to a skeptical idealism, believing in ideas, but not in things. The way to meet all this is to insist that space and time are realities such as we are led to regard them by our instinctive cognitions and beliefs.

SECTION IV.

MEMORY.

The phrase is used at some times in a wider and at other times in a more limited sense. Locke employs it to signify Retention. In our common literature it is used in a larger sense to denote all those reproductive acts implying belief. There are three of the reproductive powers implied in the exercise of memory thus understood : there is (1) The Retentive Power ; (2) The Recalling Power; and (3) The Associative Power. But the essential element is (4) The Recognitive Power. Wherever there is recognition there is memory, and wherever there is no recognition it cannot be said that there is recollection.

It is a curious circumstance that adults are not able to remember their infantine experience. But that infants remember is shown by the circumstance that they are gathering experience, for instance, learning distances and forms, which they could not do without recollection. Carpenter mentions the case of a person who remembered in after life what had passed when only a year and a half old. We may not be able to find out all the reasons of this forgetfulness of young children. It may arise from a want of tenacity in the brain, but also from the want of correlations to call up ideas. That memory fails in old age seems to arise from the want of healthy brain concurrence. It fails first in names, because they are arbitrary and have not numerous correlations to call them up.

SECTION V.

IMPROVEMENT OF THE MEMORY.

It is to be improved by taking advantage of the Laws of Association, Primary and Secondary.

We should use the Secondary Laws (see p. 135). Objects and occurrences are more apt to be remembered when they are in accordance with our Native Tastes and Inclinations. There are boys who can attain and retain a lesson in classics who cannot be made to keep hold of their mathematics, to which they have an aversion ; and *vice versa*, there are some who never forget their mathematical demonstrations who lose their classics in a short time after they have laboriously learned them. There are persons who, because of the intense interest they feel in it, can remember a hundred lines of poetry after reading them once or twice, whereas there are others on whom verse produces no impression, but who never forget the facts detailed in prose histories and books of science. In the practical professions and business of life, we find people cherishing what they have a taste for, and letting all else pass away as being utterly indifferent to them. In listening to discourses or conversation, or in reading a book, the things are apt to cling to us that have an affinity with us, and others are driven away. So far as this law is concerned we cannot directly influence our memories by an act of will; but we can do so indirectly. First, we can call forth into active operation, and we can cherish, those tastes which we wish to cultivate, by associating them with, and binding them to, ends in which we are interested. Some who have no relish for pure mathematics can be made to study them eagerly when they discover their important practical applications. Many have entered on their professional work with no great ardor for it, but they are led to pursue it eagerly as they find that it brings them wealth or reputation. Secondly, in all circumstances let us try to connect what we wish to learn and retain with some of our native inclinations. Many a boy is made to learn cheerfully an irksome task by his love to his father or his teacher. Some of us who have no pleasure in learning foreign languages have acquired them industriously, because of the treasures of literature and knowledge which are laid up in them.

We should in all cases make use of the laws of Energy. What we bestow no thought upon is sure to be forgotten ; as, having been neglected when it presented itself, it will never appear again. By turning a subject round and round in our reflections we may so make it our friend that it will visit us frequently. We may accomplish the same end by associating what we wish to recollect with feeling of some kind. By showing that we love an object, it will be encouraged to make its appearance before us. We remember the scenes of our childhood which called forth feeling pleasant or painful, whereas things unimportant, or it may be important, but which were indifferent to us, are lost forever. It is on this principle mainly that a teacher, who is beloved by his pupils and makes them feel an interest in their work, is able to impress himself and the subjects of study so deeply on their minds that they can never be effaced. But it is by the third law of energy, that of will, that we have the most effective control over our memories. We have all to regret that so much of the instruction which we received in our younger years is lost because we could not be made to attend to it. On the other hand attention puts a stamp on all to which it is strongly directed, and this gives it a continued currency.

But we may also turn the Primary Laws of Association to profitable use; as, for instance, the Law of Contiguity. We may repeat what we wish to remember, and then any part of the train, any word used to express it, will bring up the rest. Children retain for life those rhymes or passages of Scripture which they committed (to use a common but expressive phrase) to memory. If we wish to keep an object or event in perpetual remembrance let us tie it to something which is sure to come up ; say our work, or study, or devotion, to a particular place or hour of the day. If we are anxious to be reminded of a particular duty at a certain time or place, let us associate it with the persons or objects we are then and there likely to meet. We have to buy an article in a certain shop; we have a message to carry, or an intimation to make to a certain person. Let us so connect the things that when we come to the place, or meet the individual, what we wish to do is immediately suggested. We may so use the law of coexistence as to have what we wish constantly to remember, — say our business or our duties to God and our neighbor, our devotions and our alms, — associated with our habitual train, so that they come up at all times. Some have their

11

thoughts and feelings so regulated that at any time ejaculatory prayer is ready to rise to God, and their hands are ever ready to supply relief to the poor.

Thoughtful minds may also use profitably the Primary Law of Correlation. They may, as things pass under their observation, note how they are related to other things, and then these other things will recall them. If we are in the habit of noticing causes, the causes will ever after suggest the effects and the effects the causes, and we shall walk in this world as in a concatenated cosmos. Again, those who are in the habit of putting all things under heads or arranging them into classes, of course by their points of resemblance will find the law of similarity making the particulars bring up the species, and the species suggesting individuals. In the higher professions, such as law and medicine, the knowledge acquired is so assorted that it is available at all times, and comes out often in unexpected ways, on great emergencies and on small. The scientific man has his knowledge and notions arranged as in a museum, so that he can lay his hands on what he wishes at any time and place, and put every new thing that presents itself in its proper compartment. The historian lays up events under heads as the naturalist does his specimens in drawers. The very poet, though his domain is more like a garden, or a wide-spread landscape, is ever gathering up images which he is ready to plant in their proper place for æsthetic effect. A number of very eminent men intellectually are spoken of by Hamilton and others as possessing great memories, — as Julius Cæsar Scaliger, Ben Jonson, Grotius, Pascal, Leibnitz, Euler, Niebuhr, Mackintosh, Macaulay; of whom it should be observed that their memories proceeded by correlation which had been observed by their understandings.

SECTION VI.

DOES THE MEMORY DECEIVE US?

The answer is that our original and intuitive memories never do, but our acquired memories may.

The memory, using the phrase in a loose sense, does seem liable to mistakes. Two people, both honest, give somewhat different accounts of a transaction which they have both witnessed. We have all found our recollections set aside by facts well established. In order to explain the facts and save our constitution from the charge of deceit, we have to draw a distinction in regard to our memories, similar to that drawn (see above, p. 29) between our original and acquired perceptions. There are intuitive perceptions which do not and cannot err. But we are ever making additions to them by guesses and inferences meant to fill up chasms, and make our vague and confused memories clear and consistent. We have seen how much there is of inference in our ordinary sense - perceptions; we see a shape before us in a wood or in the twilight, and we conclude that it is a man or a ghost, whereas it is only a rock or a tree seen under a certain aspect. In like manner our recollection of an occurrence is dim, with breaks in it, and we proceed to fill up the figure, and make it full and consistent with itself, only, it may be, to make it inconsistent with facts.

Our original memories, having the sanction of our constitution and of God who gave it to us, seem to be confined within very stringent limits. No man can remember the whole time, and all that has occurred in it, between the present and a distant event in the past. He cannot cast a retrospective glance on the instant over the whole line between this instant and any given time mentioned, — say the time when he went to school, or began business, or when a sister died, — any more than he can tell by the eye the distance of that mountain peak.   A man is asked how many years it is since his father died.   He cannot endways see the continuous line and measure its length, and he has to inquire, to calculate, and his reckonings may be wrong.   It was before I took a particular journey, or before I married, — the eras which he regards as landmarks, which he thinks he has fixed most certainly, but which he has marked erroneously on his life chart.   Whether we are seeking to have the exact facts for ourselves, or narrate them to others, in all cases, but especially in witness-bearing, or where our words are apt to be quoted to the weal or woe of others, let us be conscientiously on our guard against going beyond our memories proper, and of adding a form or coloring which may be a perverted one, formed by the fancy under the influence of a prejudiced heart.

CHAPTER V.

SECTION I.

ITS NATURE.

IT puts in new forms and dispositions what had been previously before the mind. First, it contains a diminishing power; having seen a human being, I can picture a Lilliputian — children are greatly interested in the feats of Tom Thumb. Secondly, there is an enlarging power; having seen a man, I can imagine a giant, and be entertained with his exploits. Thirdly, there is a separating power; having seen a church, I can have an image of the steeple apart from the rest of the building. Fourthly, there is a compounding power; having seen a bull and a bird, I can put the wings of the bird on the body of the bull and fashion a winged bull such as we see on the sculptured slabs of Nineveh.

I place this faculty among the Reproductive Powers, for, far reaching as it is, it cannot produce anything of which it has not had the elements in a previous experience. Its power is always constructive, never creative. "This shows," says Locke, "man's power to be much the same in the material and intellectual worlds, the materials in both being such as he hath no power either to make or destroy." A man born blind cannot have the most distant idea of colors, nor can the man born deaf have the dimmest idea of music. But when a per-

son has seen colors, though he should afterwards like
Homer or Milton be smitten with blindness, he may be
able to combine them in unnumbered ways, all different
from that in which they are mixed in existing objects,
natural or artificial. Give one possessed of fine musical
ear a knowledge of sounds, and he may be able to dispose them so as to produce symphonies such as were
never heard before, but which, as people now listen to
them, make the soul to swell or sink with their swelling
or sinking notes.

It is the office of the memory to reproduce what has
been previously before the mind in the form in which it
first appeared, and with the belief that it has been before the mind in time past. The imagination (of which
composition is the main element) also reproduces, but it
reproduces in new forms, and is not accompanied with
any belief as to past experience. Both are reflective of
objects which have been before the mind; but the one
may be compared to the mirror, which reflects what is
before it in its proper form and color; whereas the
other may be likened to the kaleidoscope, which reflects
it in an infinite variety of new shapes and dispositions.
Each of these has its peculiar endowments by which it
is enabled to accomplish its specific end. The imagination does not, like the memory, disclose realities; but on
the other hand, the memory cannot enliven by the varied
pictures which are presented by the imagination. Each
is beautiful in its own place, provided it is kept in its
own place, and the one is not put in the room of the other;
as was said severely of an author that he resorted to his
imagination for his facts and his memory for his figures.
The one is represented by observations, experiments,
records, and annals; the other by allegories, myths,
statues, paintings, and poems. The one, as Bacon has

remarked, is peculiarly the faculty of the historian, the other of the poet and the cultivator of the fine arts.

SECTION II.

THE IMAGINATION.

There is implied in order to its exercise (1) The Retentive and (2) The Associative Power. All its images come from cognitions and ideas which have been before the mind, and are retained. They always rise up according to the Laws of Association of Ideas. The imaginations always come up according to the Laws of Contiguity and Correlation; and the peculiar character of them in the individual is mainly determined by the Secondary Laws of Native Taste and Energy.

The fancies of some follow more specially the Laws of Contiguity, and things unconnected with each other come up often to delight and amuse us by their liveliness, by their unexpected appearance, by their variety, and their curious juxtapositions. As they involve no intellectual strain we are apt to follow these in our moods of dreaminess, or when we are seeking rest and relaxation. Novels are specially fitted to gratify this propensity, and are resorted to by those who do not wish to be troubled with much thinking, and by men of business when they wish a cessation from toil. In the case of the former the constant indulgence in fiction is apt to produce a frivolous turn of mind, more and more indisposed to exertion of any kind. In the case of the latter the effect may be soothing if kept within proper limits, and the reading not carried too far into the night. But when the scenes are sensational and persons dwell often or too long among them, there may be as much wasting of the nervous power as even by business or study; and

the issue, a restlessness and dissatisfaction. The end ac-
complished by poetry, especially narrative and descrip-
tive, is much the same: to occupy the mind with exciting
images. But there is this difference between the novel,
at least the common novel, and poetry, that the latter
is usually more condensed, and therefore requires more
thought and brings before us a great many correlations
of sound and sense. The consequence is that poetry is
much better fitted than the novel to produce mental
elevation, and is less liable to the abuse of excess.

The imaginations of others are more disposed to follow the Law of Correlation, to pursue things that are
connected with each other. In our highest poetry and
poetical prose, we are called to dwell among interesting,
and it may be subtle and far ranging analogies, and among
harmonies often between material and spiritual things,
between earthly and heavenly things. Then there may
be imagination, as has often been remarked, exercised,
and this legitimately in science. In such cases the mind
proceeds according to the associative principle of correlation, and follows connections in reason, and in the
nature of things between one department of nature and
another.

But the Secondary Laws have the main influence in
determining the peculiar character of our imaginings.
Our Tastes, native or acquired, are shown as readily and
certainly by the character of our spontaneous musings as
by anything else; more so than even by our business
pursuits, which may often be determined by external circumstances. Thus we may give a direction to our fancies by associating what we wish to revive in old forms,
or in new, with exercises of intellect, of feeling, and of
will. Viewed in this light we see that we have a greater
power over our imaginations than we might at first im-

agine. What we think about and feel an interest in and attend to habitually will use the privilege of a friend and often visit us when wished for and when not wished for. In fact we can to some extent determine the character of our imaginations, good or evil, as we do those of our associates, by the friendships we form and the preferences we show.

In imagination there is

(1.) *A Picturing Power.* — A mother, let me suppose, looks out of the window of her dwelling to take one other look of a beloved son setting out to a distant land that he may there earn an honorable independence. It is a fond look which she takes, for she knows that on the most favorable supposition a long time must elapse before she can again meet with him. She continues to fix these tear-filled eyes upon him till a winding of the road takes him out of the field of view. When he has turned that corner she can no longer be said to perceive him with her bodily eyes, but the mind's eye can still contemplate him. For often, often, does she imagine to herself that scene with all its accompaniments. Often does the memory recall that son at the particular turn of the road, on a particular day, rainy or sunshiny, in a particular dress passing round that corner, and as she does so the whole is, as it were, visible before her. In this the senses are no longer exercised, but the memory, and the imagination may also begin its appropriate work. For not only will the mother recall the scene, as it occurred, — there will be times when it becomes more ideal, when one part will be separated from another, and when the parts selected for more particular contemplation will be mixed with other circumstances; and in various forms it will appear in her night dreams and reappear in her day dreams, and she will picture that son toiling and struggling in that distant land to which he has gone, rising from one step of aggrandizement to another, and returning at last by that same road and round that same corner to this same home; and she will picture herself as receiving him, not as she parted with him, with mingled fears and hopes, but with one unmingled emotion of joy, while he showers upon her a return for that affection which she so profusely lavished on him in his younger years.

(2.) *A Constructive Power.* — For the mother not only pictured the past, but put it in new shapes and combinations. Like the prisms, the imagination divides that which passes through it into rich rainbow colors.

This last is the highest property of the imagination. It is one of the characteristics of genius. It is a constituent of every kind of invention. The particular character of the invention will be determined by the native tastes and predilections, and by the acquired habits of the individual. If a person have a strong tendency to observe forms, the imagination will call up the shapes in new combinations, and if his talent is cultivated he may become a painter. If he be disposed to admire the beauties of nature, new landscapes will be apt to appear before his mind made up of dispositions of objects which he has witnessed in real scenes. When an individual has a mechanical turn, the imagination will ever be prompting him to devise some new instrument or engine; or, if his taste be architectural, new buildings will rise in vision before him. If he be a man of great flow of sensibility, he will ever be picturing himself or others — a mother, sister, or wife — in circumstances of joy or sorrow, and at times weaving an imaginary tragedy or comedy, in which he and his friends are actors.

This is a gift which like every other can be cultivated. I know, indeed, that genius is in itself a native endow-

ment. No teacher can communicate it in return for
a fee, nor can it be acquired by industry; but unless
pains be taken, it is apt to run wild and become useless
or even injurious. It admits of direction and improvement. The painter who would rise to eminence in his
art must study the finest models and fill his mind with
scenes natural and historical such as he would wish to
represent. The poet who would awaken his genius must
live and breathe and walk in the midst of objects and
incidents such as he would embody in verse. In science
discovery is commonly the reward reaped by a power of
invention which has been trained and disciplined. It is
seldom that discoveries are made by pure accident. It
was (according to the common story) on the occasion of
Newton's seeing an apple fall to the ground that the
thought flashed on him, This apple is drawn to the earth
by the same power which holds the moon in her orbit.
But how many people had seen an apple fall without
the law of universal gravitation being suggested to them!
The thought arose in a mind long trained to accurate
observation and disciplined to the discovery of mathematical relations. It was as he gathered up the frag-
ments of a crystal which had fallen from his hands to
the ground that the Abbé Haüy discovered the principles which regulate the crystallization of minerals; but
the idea occurred to one who was addicted to such investigations, and who was in fact studying forms at the
very time. On falling in with the bleached skull of a
deer in the Hartz forest, Oken exclaimed, " This is a
vertebrate column," and started those investigations
which have produced a revolution in anatomy; but the
view presented itself to one meditating on the very subject, and in a sense prepared for the discovery.

Before leaving this head it is proper to state that the

imagination can picture and put into new forms not only the material, but the mental and the spiritual worlds. The mother, in the illustration employed, can not only picture her son in new scenes, she can picture the feelings which he may be supposed to cherish in these scenes, or the feelings with which she herself may contemplate him. Milton, culling what was fairest from the landscapes and gardens which had passed under his view, describes in his Paradise Lost an Eden fairer than any scene now to be found on our globe; but as a still higher and far more successful achievement of his genius he contrives, by combining and intensifying all the evil propensities of human nature, — pride and passion, ambition and enmity to holiness, — to set before us Satan, contending with the holy angels and with God himself.

The poet, the dramatist, the novelist, dispose the elements of human nature in all sorts of new shapes and collocations, in order to please, to rouse, or instruct us. If I am not mistaken, poetry and fiction generally must be led to deal more and more, in every succeeding age, with the motives, the sentiments, and passions of mankind, — not indeed in a scientific or metaphysical manner, but in their actual concrete forms. This is a field very much overlooked by the ancients and left over to the moderns to cultivate. If we leave out of account the Book of Job and other portions of the Hebrew Scriptures, and the plays of Æschylus and other Greek dramatists, we shall find very little of the deeper moods and feelings of humanity in the poetry of the ancients. The poet who would catch the spirit of modern times must unfold the workings of the soul within as the ancients exhibited the outward incident.

I believe that the visible and tangible machinery used in times past by the poets is waxing old, and must soon vanish away. We can relish to some extent the allusion to harps and lyres; to nymphs and muses, to Minerva and Apollo, by the Greeks and Romans, for they were sincere in the use which they made of them. But it is only indicative of the barrenness of his genius to find the modern youth talking of awaking his lyre when perhaps he never saw a lyre in his life ; invoking the Muses when he believes that there are no Muses ; and appealing to Apollo when he knows full well that Apollo cannot help him. Poetry, in order to be true poetry, must come up welling from a true heart. There was nothing artificial in the use of their mythology by Greeks and Romans, but there must always be something unnatural, not to say affected, in the employment of it by the moderns. The old apparatus of the poets is now gone and gone forever, and I for one scarcely regret it. But will the scientific character of the age, which believes in astronomy and geology, and not at all in ghosts or fairies, admit of any new machinery sensible and bodily ? I doubt much if it will, for there would be no sincerity in the use of such, and sincerity must be an element in all genuine poetry.

Is the modern then precluded from the exercise of the poetic imagination ? Is the time of great poets, as some would hint, necessarily passed away ? I for one believe no such thing. But I am convinced, at the same time, that poets who would do in these times what the older poets did in their days must strike out a path different from that in which the ancients walked. The novelist has, it seems to me, already entered on this path. He has described human nature, or at least certain moods of it, — its passions, foibles, consistencies, and inconsistencies, — and so his works have had a popularity in these latter days far exceeding that of the poet. Poets are read very much in proportion as they deal with mankind. The poetry of Shakespeare ranks higher, I suspect, in this age, than that of Milton, and this mainly because the former exhibits human nature in almost every variety of attitude. Most of the greater poets of the past age delighted to daguerreotype the states of the human soul, — whether in its moods of quiet communion with nature like Wordsworth, or in the wider excursions of the imagination like Coleridge and Shelley, or in the deeper workings of passion like Byron. Even when bringing before us the objective world they often expose it to the view by a flash of light struck by the inward feeling awakened. Tennyson, in his " In Memoriam," gives us little else than the feeling of sorrow for the departed projecting itself on the external world and darkening it with its shadow.

I believe that as the world advances in education and civilization, and entertains a greater number and variety of thoughts on all subjects, and is susceptible of an ever-increasing range of emotions, poetry must take up the theme, the workings of human nature, and make this its favorite subject. This is a mine of which the ancients gathered only the surface gold, but which is open to any one who has courage and strength to penetrate into its depths and thence to draw exhaustless treasures. As the most inviting of all topics to the poet I would point to the human soul, to its convictions, its doubts, to its writhings and struggles, in boyhood and manhood, in idleness and in bustle, to its swaying motives, its desperate fights, and its crowning conquests.

SECTION III.

THE USE OF THE IMAGINATION.

The imagination has a noble purpose to serve. It widens the horizon of the mental vision. It fills the empty space which lies between the things that are seen, and it gives a peep into the void which lies beyond the visible sphere of knowledge. It thus expands the mind by expanding the boundary of thought, and by opening an ideal outside the real world. It is also fitted to extend the field of enjoyment. It peoples the waste, and supplies society in solitude; it enlarges the diminutive and elevates the low; it decorates the plain and illumines the dim. The cloud in the sky is composed of floating particles of moisture, and would be felt as dripping mist if we entered it, but how beautiful does it look when glowing with the reflected light of the setting sun! Such is the power of fancy in gilding what would otherwise be felt to be dull and disagreeable. The imagination can do more than this: it can elevate the sentiments, and the motive power of the mind, by the pictures, fairer than any realities, which it presents.

This faculty has purposes to serve even in science. "The truth is," says D'Alembert, "to the geometer who invents, imagination is not less essential than to the poet who creates." To the explorer in physical science it suggests hypotheses wherewith to explain phenomena, and which, when duly adjusted, and verified by facts, may at last be recognized as the very expression of the laws of nature. There was a fine fancy in exercise, as well as a great sagacity, when the poet Goethe discovered that all the appendages of plants — sepals, petals, stamens, and pistils — are after the leaf type, and thus laid a foundation on which scientific botany has been

built.   In every department of science this faculty
bridges over chasms between discovered truths, and
dives into depths in search for pearls, and opens mines
in which precious ores are found.

May we not go farther and affirm that it is of service
in the practical affairs of life, — always when subordi-
nated to the judgment.   Not only does it supply devices
to the inventive warrior, such as Napoleon Bonaparte,
and suggest means of reaching unknown countries to
the adventurer by sea or land : it helps the farmer to
discover new modes of tilling his land, and discloses new
openings in trade to the merchant.

Need I add that it is the power which constructs those
scenes which are embodied in the fine building or statue,
which are made visible to us on the canvas of the painter, or which the poet enshrines in verse, — as we have
seen shrubs and flowers imbedded in amber.   Generally,
those writings are the most widely diffused and universally popular which address this imaging power of the
mind.   At the head of this pictorial school is Sir Walter
Scott, and after him we have a whole host of writers in
history and in fiction.   These authors do not content
themselves with relating the bare incident : they set
before us the actors, with all their accompaniments of
locality, dress, manner, and attitude.   This pictorial
power illumines the book of knowledge, and fills it as it
were with prints and figures, which allure on the reader
from page to page, without feeling his work to be a toil.

This faculty too has the power of awakening sentiment deep and fervent.   And here it will be needful to
call attention to the circumstance that the very mental
picture or representation of certain objects — say ourselves or others in circumstances of happiness or pain —
is fitted to call forth feeling.   The novel-reader rejoices

over the success of the hero of the tale as he would over
the triumphs of a living man, and weeps over the misfortunes of the heroine as he would over a scene of actual misery.   To account for this it is alleged by some
(as by D. Stewart) that there is a momentary belief in
the reality of the object.   I am not sure that it is necessary to resort to this supposition.   It is the very mental
picture or apprehension of persons exposed to happiness
or suffering which calls forth the emotion, and this with
or without a positive belief.   No doubt if unbelief come
in it will arrest the play of fancy and feeling; and unbelief will always interpose when the picture is unlike
any reality, and hence it is needful for the novelist, the
tragedian, and the actor to make the characters and accompaniments as natural as possible, lest the doubting
judgment appear to scatter the images and with them
the emotions.   But if unbelief does not lay a cold in-
terruption on the process, it seems to me that the mental representations, as they flow on, will of themselves
draw along the corresponding train of feelings, whether
of joy or sorrow, of sympathy or indignation.

According, then, to the cherished imagination, so will
be the prevailing sentiment.   Low images will incite
mean motives, and sooner or later land the person who
indulges in them in the mire.   Lustful pictures will
foment licentious passions, which will hurry the individual, when occasion presents itself and permits, into the
commission of the deed — to be remembered ever after,
as Adam must have looked back upon the plucking of
the forbidden fruit.   Vain thoughts will raise around
the man who creates them a succession of empty shows,
in which he walks as the statues of the gods are carried
in the processions before pagan temples.   The perpetual
dwelling on our supposed merits will produce a self-

12

righteous character, and a proud and disdainful mien and address. Gloomy thoughts will give a downward bend and look, and darken with their own hue the brightest prospects which life can disclose. Envious or malignant thoughts will sour the spirit and embitter the temper, and ever prompt to words of insinuation, innuendo, or disparagement, or to deeds of sulkiness, of malignity, or revenge.

This is the darker side. On the other side, when the fancy is devoted to its intended use, it helps to cheer, to elevate, to ennoble the soul. It is in its proper exercise when it is picturing something better than we have ever yet realized, some grand ideal of excellence, and sets us forth on the attainment of it. All excellence, whether earthly or spiritual, has been attained by the mind keeping before it and dwelling upon the ideas of the great, the good, the beautiful, the grand, the perfect. The tradesmen and mechanic attain to eminence by their never allowing themselves to rest till they can produce the most finished specimens of their particular work. The painter and sculptor travel to distant lands that they may see and, as it were, fill their eye and mind with the most beautiful models of their arts. Poets have had their yet undiscovered genius awakened into life as they contemplated some of the grandest of nature's scenes ; or as they listened to the strains of other poets, the spirit of poetry has descended upon them, as the spirit of inspiration descended upon Elisha while the minstrel played before him. The soldier's spirit has been aroused, more than even by the stirring sound of the war trumpet, by the record of the courage and heroism of other warriors. The fervor of one patriot has been created as he listened to the burning words of another patriot, and many a martyr's zeal has been kindled at the funeral pile of other martyrs. In this way fathers have handed down their virtues to their children, and parents have left their offspring a better legacy in their example than in all their wealth, and those who could leave them nothing else have in this example left them the very richest legacy. In this way the good men of one age have influenced the characters of the men of another, and the deeds of those who have done great achievements have lived far longer than those who performed them, and been transmitted from one generation to another.

SECTION IV.

THE IDEA OF THE INFINITE.

The imagination is strikingly illustrative both of the strength and weakness of the human intellect. There are stringent limits laid on its exercises. All the images of the fancy are only reproductions of what we have experienced. In using its materials the mind can enlarge them to an infinite extent, but stretch itself as it may the image is still finite. In expanding the image in space it is incapable of doing more than representing to itself a volume with a distinct spherical boundary. In following its contemplation, in time the image is of a line of great length, but terminating in a point at each end. But where the mind is held in by its weakness there it exhibits its strength. It can image to itself only this bounded sphere, this line cut at both ends, but it is led, or rather impelled, to believe in vastly more. At the point where it is obliged to stop it takes a look, and that look is into infinity. Standing as it were on the shore of a vast ocean it can see only so much, but it is constrained to believe that there is a region beyond that horizon to which no limits can be set.. It is here that I find the origin and genesis of such an idea and belief as the mind can entertain of infinity.

We are approaching a profound subject. It is not easy to sound its depths. It was long before I was able to attain to anything like clear ideas on the subject. I have pondered it for successive hours only to find it shrouded in thicker clouds. On the one hand I found the more profound philosophers of the Continent of Europe, such as Anselm, Descartes, Leibnitz, and Kant, giving this idea a high, indeed the highest, place in their systems. In coming back from flights in company with these men, to inquire of Locke, Hamilton, and British philosophers what they make of this idea, I find their views meagre and unsatisfactory, for the idea of the infinite according to them is a mere impotency in respect of the mental faculty and a negation' as to the idea reached. But if we can entertain no such idea, how can we speak of it? If it be a mere impotency, how do we feel ourselves called on to clothe the Divine Being with this perfection?   '.

Feeling as if I needed to find it somewhere I proceeded in the truly British method, that is, the inductive, to inquire how does such an idea of, or belief in, the infinite as the mind can entertain rise within us, and what is its precise nature? The imagination can add and add; so far we have the large, the indefinite. Thus in respect of time (of which we have seen we have an idea by the Recognitive Power) it can add millions of years and ages to millions of years and ages. In respect of extension or space (which it knows by the senses) it can add millions and billions and trillions of leagues to millions and billions and trillions of leagues, and then multiply the results by each other millions of billions and trillions of times. But when it has finished this process it has not infinity, it has merely immensity. If, when we had gone thus far, time and space were to cease, we should still have the finite, — a very wide finite, — but not the infinite. But it is a law, and it is a conviction of the mind, that even when we have gone thus far we are necessitated to believe that to whatever other point we go there must be something beyond. Such seems to me to be the true character of the mind's conviction as to the Infinite.

The Infinite, as apprehended by man, may be regarded as having two elements, or rather may be viewed under two aspects.

I. The Infinite is always something beyond our widest image and conception. The mind strives to form an image of infinity, but as it does so it is always baffled and thrown back. It can easily picture a sphere as wide as that of the earth's movement around the sun, and try to image that vast orbit in which our sun moves. Let us stretch the imagination thus far, as far as the most distant point which the largest telescope reaches, as far as the star which requires thousands or hundreds of thousands of years to send its rays across the immeasurable regions which intervene. Are we then at the farthest limits of existence? Can we believe that we are? Suppose we were carried to such a point; would we not stretch out our hand, confidently believing that there is a space beyond, or if our hand be hindered, it must be by a body occupying space? We are necessitated to believe that when we have gone thus far we are not at the outer edge of the universe of being; nay, though we were to multiply this distance by itself ten thousand millions of times, till the imagination feels dizzy and reeling, still, after we have reached that point, we are constrained to believe that there must be something beyond. This seems to me to be the law of the mind in reference to infinity; it not only cannot set limits to existence, it is constrained to believe that there are no limits. "If the mind," says John Foster, "were to arrive at the solemn ridge of mountains which we may fancy to bound creation, it would eagerly ask, Why no farther? What is beyond?"

II. The Infinite is such that nothing can be added to it. We may farther say that nothing can be taken from it. It is THE PERFECT.[1]

All that we know by our highest faculties and in our most elevated moods is seen to be limited, and in this sense, and it may be others, it is imperfect. But amid all the excellencies and evils before it it is ever looking out for that which has no deficiencies. So whatever is known to us as great and good we stretch to the uttermost, and combine all in one; and would unite Almighty Power, Omnipresence, Eternity, Omniscience, Boundless Goodness, and Spotless Holiness all in this Perfect One. The mind is made to acknowledge that it cannot compass all this, but is expanded in the endeavor to comprehend it. The imagination loses itself as in a forest; but we feel all the while that we are safe, wherever we are, — in the immensity of space or time or eternity, in this world or in worlds unknown. I have been speaking of our rudimentary faith in the unseen and the distant; we have now come to a faith in what cannot be transcended. We have now a grand ideal set before us to contemplate, and though like the pole-star it is far above us, it is there to guide us. We are ever drawn towards it, and as the asymptotes of the hyper-

[1] After working out this twofold aspect I found that I had been antici pated by Aristotle. See *Intuitions of the Mind*, Part ii. B. ii.

bola ever draw nearer, while they never touch each other, so we would ever approach that model which is yet ever above us.

This second aspect of infinity is the grander and the more important. This was the feature brought into prominence by Anselm, the great mediæval philosopher and theologian. It was the one fixed on by Descartes, the founder of the French philosophy, and by Leibnitz, the originator of the German philosophy. We find the germ of it, ready to be expanded, in the minds of all men, if we go sufficiently far down. We strike upon it in all our deeper reasonings in regard to Divine things. The profound philosophers just named argued from the very existence of such an idea in the soul, that there must be a corresponding object, and that therefore God exists. Whatever may be thought of the validity of this argument it is certain that there is such a rudimentary idea in the mind, and that it is ever prompting us to seek after God, and enabling and constraining us when we get evidence of the existence of God, say from the traces of design in nature, to clothe him with infinity.

This second aspect of the infinite is not inconsistent with the first, but is complementary to it. Combine the two and we have such a view as man can entertain of the infinite. By the one aspect he is humbled under a sense of inferiority; by the other he is elevated as he gazes on it. Certainly, man's idea of the infinite is not an adequate one; he is made to feel so as he entertains it. But it is not a negative idea, or a mere impotency, as Locke and the British school of philosophy hold; it has positive elements in it, and man is never more exalted than when he is seeking to rise to it. The belief may be regarded as an intuitive one; in our deeper moods we find ourselves gazing on it. It is a necessary one; we cannot be made to think otherwise. It is, in a sense, a universal conviction. No doubt the widest image formed by human beings, as by children and savages, must be very confined ; but, narrow or wide, we feel that there must always be something beyond. Pursue any line sufficiently far and we find it going out into infinity. So true is it, as Shelley says, —

> " The feeling of the boundless bounds
> All feeling, as the welkin doth the world."

But the infinite which the mind is led to believe in is not an abstraction. It is a belief in something infinite. So when "the visible things of God" declare that there is an intelligent being, the author of all the order and purpose in the universe, the mind is constrained to believe that he is infinite, and clothes him with "eternal power and godhead."

SECTION V.

THE ABUSE OF IMAGINATION.

While the imagination is fitted, when properly regulated, to widen the field of enjoyment and elevate the standard of character, there is no faculty which is more liable to run into error and excess, and in the end to land the possessor in more helpless and hopeless misery. If I had the genius of Plato, and were able like him to clothe my thoughts in instructive myths, I would represent the God who created us as allotting, when he distributed to the faculties their proper spheres of dominion, to the understanding the land, to the passions the sea, and to the imagination the air. While each has a kingdom put under it, it is all the while under a higher Sovereign to whom it must give account, and who is ready to punish if his eternal laws are contravened.

And there may be transgression, not only in erroneous judgments, not only in violent passions, but in the imagination wandering into forbidden regions. No sin brings its punishment with it more certainly in this life than a disordered imagination. This kingdom of the air, just as much as the land or the sea, has had laws impressed on it. If the land is not properly cultivated it will yield no crops; if the sea is not skilfully navigated it will speedily dash the vessel in pieces ; but the air is, if possible, a still more perilous element to wield than the earth or the ocean, and the penalties which it inflicts are still more fearful; when it is offended it raves in the storm, it mutters in the thunder, it strikes with its lightning. How melancholy have been the lives of very many of those who have possessed in a high degree that fearful gift, the gift of genius! One who was himself possessed of high genius was wont to thank God, because he could discover no traces of poetical talent in his son ; and when we read the lives of the poets we can understand how Sir Walter Scott—for it is to him I refer—should have felt in this way. For in how many cases has their elevation above other men been like that of Icarus: they have mounted into a region purer and more fervent than this cold earth, only to find their wings melted by the heat, and their flight followed by a melancholy fall. This is a gift which young men of noble aspirations are especially apt to covet, and if they possess the gift by all means let them use it; if God has given them wings let them soar. But let them know that if the gift is abused, in very proportion to the greatness of the endowment will be the greatness of the punishment. For in this unreal world of their own creation, they will meet with horrid ghosts and spectres (also of their own creation, but not on that account the less dreadful), ready to inflict vengeance upon those who have made an unhallowed entrance into forbidden regions. The miseries of men of genius have been the deepest of all miseries, for the imagination has intensified all the real evils which they suffer, and added many others, giving a greater blackness to the darkness in which they are enveloped, and a keener edge to the weapons by which they are assailed.

The youthful mind, especially if of a vain or of a pensive and indolent turn, is much tempted to exercise the imagination in castle building. Speaking of his younger years, Sir James Mackintosh tells us: "Reading of Echard's Roman History led me into a ridiculous habit from which I shall never be totally free. I used to fancy myself Emperor of Constantinople. I distributed offices and provinces among my school fellows. I loaded my favorites with dignity and power, and I often made the objects of my dislike feel the weight of my imperial resentment. I carried on the series of political events in solitude for several hours. I resumed them and continued them from day to day for months. Ever since I have been more prone to building castles in the air than most others. My castle building has always been of a singular kind. It was not the anticipation of a sanguine disposition expecting extraordinary success in its pursuits. My disposition is not sanguine, and my visions have generally regarded things as much unconnected with my ordinary pursuits and as little to be expected as the crown of Constantinople at the school of Fortrose. These fancies indeed have never amounted to conviction, or, in other words, they have never influenced my action, but I must confess they have often been as steady and of as regular occurrence as conviction itself, and that they have sometimes created a little faint expectation, or state of mind, in which my wonder that they should be realized would not be so great as it naturally ought to be." A person of a very different temperament, Charlotte Elizabeth, describes herself as falling, in her younger years, into a similar habit, which, however, she speedily corrected. " I acquired that habit of dreamy excursiveness into imaginary scenes and among unreal personages, which is alike inimical to rational pursuits and opposed to spiritual-mindedness." I have remarked in my own experience (for I confess to have been an architect of these airy fabrics) that all such "vain thoughts" sooner or later end in sadness ; — after the height comes the hollow, deep in proportion to the previous elevation ; after the flow comes the ebb to leave us stranded on a very sandy waste. The mind, when it awakes as it must, revenges itself for the dreams by which it has been deceived. For the time they enfeeble the will, they relax the resolution, they dissipate the energies, and they issue in chagrin, disappointment with the world, ennui, and not unfrequently bitterness of spirit. The indulgence in such weak imaginations is like the sultry heat of a summer day : it is close and disagreeable at the time, and it is ever liable to be broken in upon by thunders and lightnings. These gathering clouds, though they may seem light and floating, will sooner or later pour forth tempests. They that sow the wind shall reap the whirlwind. If the imagination is unlawfully engaged when building palaces among the gilded clouds, it is equally misemployed when, under the guidance of a melancholy spirit, it is hewing out sepulchres in desolate and gloomy places, and peopling them with ghosts and demons to keep the timid from going out into the dark night when duty calls. "Sufficient unto the day is the evil thereof."

This vain spirit is much fostered and increased by the excessive novel-reading of the age. I am not to enter upon a crusade against the perusal of works of fiction. I should be sorry to debar the child from Robinson Crusoe or the Pilgrim's Progress, or to prevent any one from becoming acquainted with the character of Jeanie Deans or of Uncle Tom. But I do protest against that constant and indiscriminate perusal of romances in which so many indulge. In the use of such stimulants I am an advocate, not of total abstinence, but of temperance principles. I am not afraid of an occasional stimulant, provided people be not constantly drinking of it, and provided they be taking solid food in far larger measure. For every novel devoured let there be eaten and digested several books of history or of biography, several books of voyages and travels, several books of good theology, with at least a book or two of science or of philosophy. If you examine some of our circulating libraries you will find a very different proportion, — far more works of fiction than of truth. Those who consume this garbage will soon take its hue, — as the worm takes the color of the green herbage on which it feeds; and the furnishing of their mind becomes excessively like the circulating libraries to which I have referred, — a strange medley, in which the vain and fictitious occupies a far larger place than the real and the solid.

Nor let it be urged by the novel-reader that as he does not believe the tale when he reads it, so no evil can possibly arise from the perusal of it. For the mischief may be produced altogether independent of his belief or his disbelief. It arises from the impressions produced, unconsciously produced, unconsciously abiding, and unconsciously operating. Like the poison caught from visiting an infected district, it is drawn into the system without our being aware of the precise spot from which it comes, or even of its existence. Like the evil influence of companions, these evil communications corrupt good manners, all the more certainly because they work pleasantly and imperceptibly. The evil arises from the vain shows into which the mind is conducted; from the false pictures of the world and of human character which are exhibited. It springs from the images with which the mind is filled, and which present themselves when invited and when not invited. For having called up these spirits, and cherished and fondled them, we may find that we cannot lay them when we choose; that they abide with us whether we will or no, first to tempt and finally to torment us.

Even when the novels are all proper in themselves the immoderate use of them has a pernicious tendency. It has been shown by Bishop Butler and by Dugald Stewart [1] that it is injurious to the mind to stimulate high feeling, — as is done in the novel, — when the feeling is not allowed to go out in action. It is a good thing to cherish compassion towards a person in distress, when we are led in consequence to take steps towards his relief. But it is not so good a thing to indulge in sympathy towards an imaginary personage whom we cannot aid. The rationale of this can be given. In proportion as we become familiar with scenes of distress we are less and less affected by them. But when the scenes are real, and when we are in the way of relieving the misery, we are in the mean time acquiring a habit of benevolence, which like other habits will grow and strengthen with the exercise. In going into such scenes we may not feel so keenly as we at one time did, but if the mere sensibility of benevolence is lessened, the principle and the habit

[1] Butler's *Analogy*; Stewart's *Elements*, Part i. cap. viii.

are increased. But it is different when our feelings are in the way of being roused by harrowing scenes in a romance; here we have the feelings deadened to ordinary misery without any habit of active benevolence being acquired. Hence it is that we so often find that the eyes which stain the novel with tears refuse to weep over the real miseries of the poor. " From these reasonings it appears," says the philosopher last named, " that an habitual attention to exhibitions of fictitious distress is in every view calculated to check our moral improvement. It diminishes that uneasiness which we feel at the sight of distress, and which prompts us to relieve it. It strengthens that disgust which the loathsome concomitants of distress excite in the mind, and which prompts us to avoid the sight of misery, while at the same time it has no tendency to confirm those habits of active benevolence without which the best dispositions are useless."

This is the result even on the supposition that the characters are properly drawn. Still more fatal consequences follow when the imagination is employed in such works to decorate vice or depreciate true excellence; to picture human nature as essentially good and the ungodly as truly happy; to represent piety as mean or profanity as something noble; to picture the religious as either fools or hypocrites; or daub over with paint the face of fading worldly vanity.

SECTION VI.

TRAINING OF THE IMAGINATION.

It may best be educated by laying up a store of noble images, ever presenting themselves to enliven and instruct the mind. There are works devised by the imagination of man fitted to accomplish this end. There is the statue with the soul shining through the marble. There is the painting, setting before us historical incident and character, and rousing the soul to high sentiment and energetic action. There is the grand cathedral with its imposing towers, its pillar succeeding pillar, and arch upon arch, with the long prospective of the nave and the withdrawing aisles. It is worth our while to travel many a mile to store the mind with such memories.

But the works of God are still more replete than those of man with food for the fancy. Nature everywhere brings before us figures which strike the eye, which imprint themselves on the memory and engage the musing intellect. The planet has a regular oblate spheroid shape, and it runs in a regular elliptic orbit. Minerals assume crystalline forms which are mathematically exact. The mountains stand so stable and leave their figure on our mind so distinctly as they cut the sky. But it is in organic nature that type has most significance. The elementary form is the cell; then there is what I call the organic column, being a shaft widened at the two ends, seen in the stalks of the leaf, in the boles of trees, in the fingers, and in all the bones. All the parts of a flower are formed on the model of the leaf, and I have shown that there is a correspondence between the form of the leaf and the form of the branch and of the whole plant. How beautiful an object is a tree growing fully in a sheltered lawn; how picturesque the same tree in winter, so sharply defined by a frost-bound covering of snow. Now the fancy is interested, and through it the meditative intellect, when " man in his spirit communes with the forms of nature."

No one has traveled much among the lovelier or grander of nature's landscapes without witnessing scenes which can never be effaced from the tablet of the memory, but which are photographed there as by a sunbeam process. It is a quiet valley separated from all the rest of the world, and in which repose visibly dwells. Or it may be a wide extended plain and fields with hedge-rows and scattered trees, and dotted over with well-fed kine which need only to bend their necks to find the herbage ready to meet them; and a river winding slowly through the midst of it, with villages and village churches on either bank, — the church towers fixing the whole scene in the memory. The ship with its pointed masts and its white sails stretched out to the breeze makes the bay on which it sails lively and attractive. More imposing are the bold mountains which cleave the sky, and the scarworn rocks which have faced a thousand storms and are as defiant as ever. How placid is the lake sleeping in the midst of them, sheltered by their overhanging eminences and guarded by their turreted towers; heaven above looks down on it with a smile, and is seen reflected from its bosom. Grander still, there is the ocean, always old and yet ever new in its aspects, never changing and yet ever changing; and the sea-bird, careering from cliff to cliff, and hoarsely chiding all human intruders from what it reckons as its own domains. The faculty which God has given us is educated by the contemplation of the scenes which He has placed around us. A stroll among such scenes at least once a year, when our large cities give clouds of dust but refuse to give us breath, is as exhilarating to the mind as it is to the body; and the mental vigor resulting will continue longer than the revived bodily vigor; while the pictures hung round the chambers of the mind will be seen looking down upon us ever and anon, to relieve the irksomeness of our daily solicitudes.

But human nature, with its joys and sorrows, its achievements and disappointments, is better fitted to stir up our higher faculties than the grandest objects fashioned out of matter. History and biography reveal incidents which incite the imagination, and youth should be made acquainted with them. They bring under our notice characters which transcend in grandeur the greatest of the works of nature, — its mountains and its vales, its streams, its cataracts, and its precipices. Those who would train the mind to its highest capacity must furnish to the young the record of deeds of heroism, of benevolence, of self-sacrifice, of courage to resist the evil and maintain the good. Friendship, fidelity, patriotism, and piety must be presented in their most attractive forms. It will be acknowledged, even by those who fail to discover that the Scriptures are inspired, that they bring before us the incidents best fitted to interest the young and to improve the character.

I have been uttering a word against the excessive novel-reading of the age. But works of fiction in poetry and prose gratify the powers which God has given us, and if consistent with moral and religious principle may refine and enlarge them. Let us look to the highest models which the highest writers of ancient and modern times have set before us. It appears to me that some of the tales presented to the young, some even of our Sabbath-school stories, tend rather to dissipate and weaken the mind. Others give utterly perverse views of human life, and make men, women, and children act from motives which never swayed human beings. Let us not confuse the mind by having presented to it a multitude of fictitious scenes, which tend to efface each other. But let us have a limited number of images stored up, each standing out prominently and distinctly. Let these

13

be the characters which have become classical by being represented by the great writers of ancient and modern times.

By all means let the minds of youth be inspired by tales of heroism. But let me not be misunderstood. I do not regard that man as a hero who has slain hundreds of thousands of his fellow-men, but who has been all the while the slave of his own ambition. I trust that as the world grows older it will also become wiser, and reserve its admiration for men of a higher stamp. By heroes I mean those who have risen above the meanness of the world, above their age, it may be above themselves, who have sacrificed their own interests to the good of others, who have aimed at nothing less than rendering their fellow-men wiser and better. A heroism, this, to be found as readily in the cottage as in the palace; in the cabin among the mountains or the most obscure alley of a great city as in the camp or battle-field; in the weaker woman as in the stronger man. She is a heroine in my estimation who, knowing that she risks her life, nurses night and day the brother or sister who is in raging fever and breathing infection all around. He is the hero who, in the midst of pollution, temptation, and defalcation, holds himself high above them and refuses to be contaminated. Every one may claim a noble lineage who is sprung from ancestors who displayed such qualities. He is of no mean descent who can claim an honest father and a virtuous mother. A man's personal experience is valuable in proportion as it has brought him in contact with persons of high soul and noble aims. Highly privileged is the youth who has had a father who has set him an elevated example, or a mother who forgot herself in attending him, who has an attached brother or sister, or who has gained a disinterested friend, willing to stand by him in misfortune. There is a sort of education which ennobles a youth more than book or training in school or college. These home scenes are more instructive than foreign travel of any description. The image of such a sister, of such a wife, is more pleasing and benign than the recollection of a painting of a Venus or Madonna. The remembrance of a friend who defended us is more invigorating than that of a statue of a Hercules or Apollo. A man whose mind is stored with these memories is never alone, for he has friends to travel with him wherever he goes, to enlighten him with their wisdom, and warm him with their love. By means of such scenes the imagination is inspired; and out of them it constructs its cherished fancies and its ideal world.

CHAPTER VI.

THE SYMBOLIC POWER.

SECTION I.

ITS NATURE.

By this is meant the power of thinking by means of signs or symbols, especially Language.

When the objects are now present and under the senses external and internal, such as the mountain before me and the joy we feel in contemplating it, we do not need any sign to enable us to think of them. We can compare these two statues without the use of words or any other medium, and decide for ourselves that the one has better proportioned form and the other has more intelligence and expression in the countenance.

When the objects are absent we need ideas of them in the mind, what I call phantasms, in order to think of them. By means of these we can compare two statues not before our eyes. In such cases we primarily compare the images, but these images stand for the things, say the two statues, and we regard ourselves as comparing the things as imaged.

When the ideas or images are singular we can easily think of them ; we can compare them and reason about them by means of the ideas which represent them. It is thus we can discover resemblances and contrasts between Homer and Virgil, between Alexander the Great and Julius Cæsar, Newton and Leibnitz, Locke and Kant, Washington and Abraham Lincoln, Napoleon Bonaparte and Louis Napoleon, the English, American, and French Revolutions.

Even when we are thinking of qualities or classes (abstract or general notions), we can do so by means of phantasms. We have occasion, let me suppose, to think of elasticity (an abstract term), and we image a rubber ball, which can be squeezed, but speedily takes its original shape. We have to reason about roses (a general notion), and we make an effort to place before the mind a plant which has all the qualities of . the rose without those of other plants, such as the daisy or lily. This kind of idea is much dwelt on by Locke. "Thus in forming our idea of man we leave out of the complex idea that which is peculiar to the individuals, — that which is peculiar to Peter and James, Mary and Jane, — and retain only what is common to all." (Essay, B. III. 3.) Locke is certainly right in holding that we do endeavor to form such an idea. Farther, we do in fact think, compare, and judge, by means of such phantasms. But it should be observed that these are not the same as our abstract and general notions. They are concrete and singular, and so are not the same as abstract notions, which are notions of parts or attributes of objects, or as general notions of an indefinite number of objects joined by the possession of a common attribute. In all such cases the phantasm is used as a sign, the individual image standing for a quality which it strikingly exhibits, or for a general notion of a member of which it is a sign. We can often think correctly enough by means of such ideas. When we can do so we move in the midst of pictures, and our thinking is rendered livelier and more interesting.

Such ideas, it should be noticed, are always inadequate when considered as standing for abstract and general notions. They are concrete, and present more than the abstract idea; they picture the object as well as the attribute. Bishop Berkeley exposes with great acuteness the absurdity implied in the supposition that the mind can form abstract general ideas in the sense of positive representations. "The mind having observed that Peter, James, and John resemble each other in certain common agreements of shape and other qualities, leaves out of the complex or compounded idea of Peter, James, and any other particular man that which is peculiar to each, retaining only what is common to all, and so makes an abstract wherein all the particulars equally partake, abstracting from and cutting off all those circumstances and differences which might determine it to any particular existence. And after this manner, it is said, we come by the abstract idea of man; or, if you please, humanity or human nature, wherein it is true there is included color, because there is no man but has some color; but then it can be neither black nor any other particular color wherein all men partake. So, likewise, there is included stature; but then it is neither tall stature nor low stature, but something abstracted from all these." (Int. to Prin.) Such considerations show conclusively that the mind cannot form any just or adequate idea in the sense of image or phantasm of a class. The truth is that every image before the mind must be that of an individual, and cannot therefore fully exhibit a species or a genus. And as the notion becomes more and more general or more abstract, especially when it is of mental or spiritual objects, the representation becomes more and more difficult; and in our higher intellectual processes it is felt to be impossible to form any picture. Who can form an idea, in the sense of image, of gravitation, of law of virtue, of expectation, of indignation, of civilization, of government?

It is a mistake to suppose that we cannot think, cannot compare, or reason, or feel, or approve, or disapprove without language. It is necessary that man should first think before he can understand language. Certainly it must have been necessary for him to judge and reason before he could invent language. Man thinks primarily by means of phantasms, and these can be made to stand for higher thoughts and used accordingly.

Man uses signs in the first instance to indicate his wants, and to express his meaning to others. Even the lower animals, such as crows, have the capacity of using signs to announce the discovery of food or of danger to their species, and of understanding them. Laura Bridgman, without the senses of sight or hearing, had a disposition to resort to movements of the body to express her thoughts and feelings. Children are commonly disposed to ring their vocables the livelong day. Homer gives it as one of the characteristics of mankind that they are word-dividing ($\mu\acute{\epsilon}\rho o\pi\epsilon s$), analyzing and constructing to form a language suited to their ever-advancing thoughts, and using that language to advance their thoughts still farther.

In our higher abstractions and generalizations, and in our reasonings and moral judgments, we need symbols, and especially language, to carry on our mental processes. They are as much required as figures are in arithmetic, as letters a, b, x, y, in algebra. In the first place, it would be difficult to form a mental image of 10, of 856,678, or of $\frac{a}{b^2c^3}$, or of personal prejudice, or benevolence, or self-sacrifice, or spiritual purity, or perfection. In the second place, these phantasms in our uneducated reasonings and recondite researches might have a confusing, a distracting, and misleading influence, as bringing objects and qualities not relevant, or omitting qualities essential to the argument.

In feeling his need of them and finding the use of them, man comes to carry on his thinking, to a great extent, by means of language. In this way his thinking is abbreviated, by using simple words for very complex thoughts, and can be carried on more rapidly and much farther.

It should be adopted as a principle, however, that in thus using signs for thoughts we should always be ready to translate the sign into the thing signified. In discussion an opponent is entitled to insist on this. In recondite reasoning, in which confusion is apt to appear, we should do it for our own satisfaction, lest we be led to affirm or deny of the sign what we would never predicate of the thought or thing for which it stands.

I am inclined to give the Symbolic a place among the faculties of the mind. It may be difficult to determine how much of it we owe to original capacity and how much to development and heredity. It seems to me to be the product of a combination of several powers. There is in man an organic apparatus of a very flexible character, and capable of producing a greater number and variety and delicacy of tones than any artificial instrument; it consists of the larynx and its attached organs, the epiglottis, ligaments, and chords. The power of association of ideas is always involved in it; the thought is associated with the sign. However produced, Language is to man a natural endowment, and is to be regarded as a heaven-bestowed gift.

SECTION II.

RELATION OF SPEECH TO THE BRAIN.

[I am indebted to my former pupil, M. Allen Starr, M. D., Ph. D., for the statement in this Section. It is altogether worthy of being recorded.] While there seem to be local centres in the brain, there is at the same time a unity of brain action. Since speech is the embodiment of a mental act in physical vibrations, it is evident that it has both material and psychical elements. Each word which is intelligently used has been acquired by a process of education. If that process be analyzed, it may be shown that the mental basis of speech consists of a series of word-images, each made up of a number of memories. Thus the word *bell* has its mental elements, which may be distinguished from one another as follows, and illustrated by the aid of a diagram: There is the memory of the sight of the bell, which may be called the visual memory. There is another of its tone, which may be termed the auditory memory. There is one of touch, which recalls the rough cold surface of the metal, the tactile memory. Then the word *bell* as heard differs from the tone of the bell, and is preserved in the word-hearing memory. Also the word *bell* as printed or written must have been retained in the word-seeing memory. Finally, there are two effort memories connected with the muscular movements involved in uttering and in writing the word. Thus the word-concept *bell* is made up of a number of memories, each of which in the diagram is represented by a circle. The various memories are, however, associated intimately with one another, so that when one is aroused the others come to mind. The circles must therefore be joined with one another by lines in the diagram, which then represents the mental elements of the word *bell.* The physical elements may now be considered. Each of the memory-pictures of the bell is the relic of a past perception, which has been acquired through an organ of sense. The visual memory is the reproduction of a perception of sight obtained through the eye; the auditory memory is the recollection of a tone heard by the ear; and so for the other memories. Each organ of sense is a physical mechanism capable of receiving vibrations, and is connected by a nerve with its own region on the gray surface of the brain, to which the vibrations are sent as sensations, and in which they are perceived. In the diagram we may therefore join the circles with the bell by lines, which will represent the nerves from the organs of sense, and to the organs of motion.

If the memory-picture is a relic of a perception, it follows that the memory is located in the same region in which the perception occurred. But anatomy has shown that the various regions of the brain are joined with one another by nerves which run beneath the gray surface in the white matter; so that the lines joining the circles in the diagram may represent the association fibres of the brain as well as the mental connections of the memories. The diagram is therefore more in accordance with an actual arrangement in the brain than it may have seemed at first.

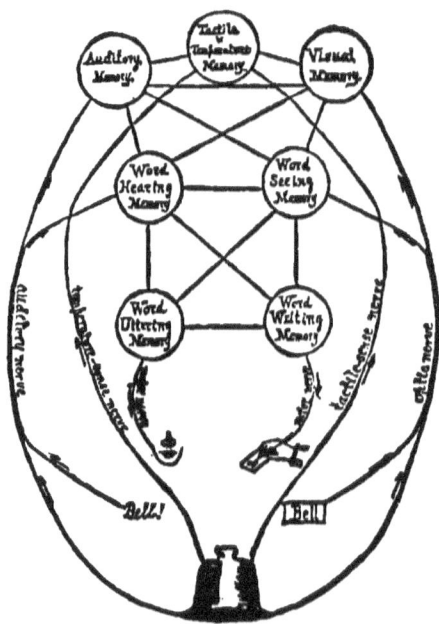

<small>DIAGRAM OF THE WORD-IMAGE "BELL." (Modified from Charcot.) Each circle represents a distinct memory involved in the mental image. The circles are joined together because the memories are associated in the mind. Each memory is the relic of a past perception, acquired through an organ of sense. The lines to the circles indicate the source of the perception. The organs of motion by which the word is spoken or written are the mouth and the hand.</small>

If, now, I show you a bell, and ask you its name, your visual memory is first aroused, then your word-hearing memory, and, finally, your word-uttering memory; three distinct memory-pictures rising in your mind in succession by the process of association. If, however, I merely ask you to repeat the word *bell* after me, I arouse but two memories in succession ; one the word-hearing, the other the word-uttering. The latter, being a simpler process than the former, is found by actual measurement to require but one half of the time ; for the repetition of a word takes but one fourth of a second, while the naming of an object takes about half a second.

But if the memory pictures are really distinct from one another, and lie in different regions of the brain, it should be possible for disease limited to one region to produce a loss of one kind of memory. And this is actually the case. For it is found that some persons lose their memory of objects, so that they do not recognize them when seen; others lose the power of understanding spoken language; others forget how to read or write; and others, still, lose the power of speaking the words which they know and remember. So that there are diseases of the brain whose effect is to deprive the person suffering of a single set of memory-pictures, an effect which could be represented on the diagram by obliterating one of the memory-circles. And further, there are forms of disease which affect the association-fibres joining different areas of the brain, in which case the association of ideas is interfered with. If, however, the surface of the brain in these cases is not destroyed the memories remain, and it is often curious to see the manner in which they are reached by indirect association when the direct fibres are broken. All these facts point to the existence of a physical basis of speech in the brain, which corresponds, as we have seen, quite closely to the mental basis. It is an interesting fact that only one hemisphere of the brain presides over the process of speech : in right-handed persons it is in the left hemisphere that the memories are stored; while in left-handed persons it is the right hemisphere which preserves the mechanism. The study of diseases of memory has led to the discovery of the facts mentioned, and is likely to throw much light on other mental processes now imperfectly understood.

SECTION III.

ON THE TEACHING OF LANGUAGES.

From an early age children are very much dependent on symbols (as all teachers know), and especially on language, for the exercise of thinking. To a small extent this may be a disadvantage, as in the use of words they are made to think as those do who have coined the phrases and who use them. But to a far larger extent it is a benefit, as it puts them in possession at once of the matured thought of ages. The power of speech, early practiced and going down by heredity, is a natural endowment and should be cultivated by children. I rather think, however, that young children should not be distracted by learning any other tongue than their own, which they should be taught to use correctly. But great advantages arise from people who claim to be educated being instructed in other tongues as well as their own, as they are thereby introduced to the thoughts of other peoples, and are not bound to move in the ruts which have been worn by their countrymen.

A talent for languages is developed at an earlier age than one for mathematics or physics. At the age of nine or so a child may begin to learn Latin or French, but should not be pushed hard. In a year or two afterwards Greek or German may be added, great care being taken not to overload the brain or to confuse the thinking. I find that a much greater number of young people from twelve to sixteen or so betake themselves with more eagerness to languages than to abstract science ; advantage should be taken of this taste to have the teaching of languages commenced in childhood, and I am disposed to add completed in youth, except, indeed, when linguistic scholarship is sought, when it may have to be continued for life.

If we wish to make the acquisition of foreign languages attractive they should be learned in much the same way as our native tongue has been. There must indeed be simple grammatical rules, gathered from the passages read, taught from the beginning. But the more scientific grammatical and linguistic laws should not be insisted on till the scientific faculties have been so far matured and are ready to work.

I am bound to add that when the sole education or even the main part of it has been in languages, the training is not favorable to independence or to solidity and manliness of thinking. When children rise to fourteen years or so, scientific should be mixed with the linguistic studies, if the mind is to be fully or healthily developed.

SECTION IV.

THE TRAINING OF THE REPRODUCTIVE POWERS.

I have taken pains, in my exposition of the separate powers, to show how they may be cultivated. It now remains only to gather the remarks to a point. The reproductive powers come next to the senses and the accompanying consciousness, in the order of their appearance. They should be educated in early life, in order to call forth and prepare materials for the higher powers, such as the judgment and the conscience. Exercise in memory and in language, if we follow the course of nature, should come before science, into which, as I think, some modern educators would hurry children at too early an age. Our fathers were right in exercising the memory, the apprehension, and fancy before introducing youths to the more abstract problems of science ; but they often erred in burdening the mind with too many dry details, or in engrossing it with words.

(1.) Pains should be taken to retain knowledge and all useful lessons in the mind. I have carefully explained how all this may be done. In order to this it is needful that the teaching should be as interesting as possible, —

that it should engage the intellect by everything being explained and the attention being thoroughly secured.

(2.) It is of vast moment that the association of thoughts and feelings be properly regulated, that vice be not painted as something grand and noble, and virtue as something mean. We must not be satisfied to have youth learn by rote, that is, by the mere law of contiguity; they must lay up facts in classes and according to the relations of causes and consequences.

(3.) I have shown how the memory may be improved by taking advantage of the laws of association, primary and secondary. Particular pains should be taken to make children distinguish between the original and proper memories and the color which may be given and the additions made by association and by rapid inference. It is thus that we can have truth without a mixture of fiction, and, what is one of the most valuable of virtues, a spirit of truthfulness.

(4.) A stock of images, pure, chaste, and ennobling, should be laid up in childhood and in youth, to be called up in after years in the midst of the cares of business and the lassitude of infirmity. Education should not be made too mechanical or even scientific. Children should be induced to read tales of heroism and magnanimity, to watch the aspects of nature, and to mingle in scenes full of human interest.

(5.) Our forefathers in some schools gave too exclusive a place to language. But it is certain all the while that language is a natural gift, that children can learn a new tongue before they can learn a science, and that languages, especially our own language, should be cultivated from an early age, for the training they give and for the knowledge they open to us.

There is a keen dispute in the present day as to

whether language and literature or science should hold the higher place in our institutions of learning. If we are to look to the place which God has assigned to these two departments, we should give to each an equally important position, and not forget to complete the trinity by adding philosophy or the branches which inquire into the foundation of knowledge and the reasons of things, and call forth the powers of thought and reflection.

BOOK III.

CHAPTER I.

OFFICE OF THE COMPARATIVE POWERS.

HITHERTO every mental perception or apprehension coming before us has been singular. All objects observed by the senses, external and internal, have been unconnected. These, when reproduced by the memory, and even by the imagination, are still units. By the latter of these powers we may join the tail of a fish to the body of a woman, but the mermaid thus fashioned is quite as individual a thing as the woman or the fish in our idea of it. We are now to consider the mental power which notices the relations of objects and thus binds them in our apprehension. It may be called Comparison, and is defined as the Faculty which discovers Relations. It observes, first, the relations of objects given by the simple Cognitive and Reproductive Powers, and then goes on to observe relations between these, on and on to an indefinite extent; it can notice the relation of classes to classes, and pursue effect on to cause, and a cause on to a prior cause, and so with all other relations.

(1.) The discovery of relations proceeds on a knowledge of the objects related. Even as the objects perceived to be related are real, so are also the relations perceived. I lay down this proposition in opposition to one of the skeptical doctrines of the present day. There are metaphysicians who tell us that things themselves are unknown to us, and that we perceive only the relations of things. This makes the relations perceived subjective, that is, merely in the mind. In standing up for the veracity of our cognitive faculties and the reality of things, we should set aside both these positions and maintain that the things perceived and the relations perceived between them are both real. No doubt the reality of the two is somewhat different : the reality of substances and the reality of the relation of substances. But as the substances — say mind and body — exist, so do the relations exist in the substances. These two lilies exist, but so also does their resemblance in the possession of the same form. The things that constitute a cause are real, and also those which constitute the effect, but the power in the cause to produce the effect is also and equally a reality.

(2.) Man's knowledge begins, not with relations, but with things. In laying down this proposition I undermine one of the most fatal — as I regard it — errors of the day. It is said that all man's knowledge is relative. I look on this as a mistake, logically and chronologically. Our consciousness being witness we regard ourselves as having a knowledge of this thing and that thing, — say of a brown horse, or of ourselves as perceiving it. Having got this knowledge, we may then compare two or more objects thus known and discover some connection between them, — as that they are like each other, or that they differ from each other, or that the one attracts the other. We discover the relation because we so far know the things and perceive the relations to be in the things.

14

This gives us a positive, as opposed to a relative, theory of knowledge. Instead of saying that we know the relations of things themselves unknown, the correct statement is that we discover the relations of things known, and discover the relations because we know the things. In this way we avoid that most subtle skepticism of our day, which begins with the doctrine of Relativity and ends with Nescience or Agnosticism.

(3.) It is wrong to maintain, as so many do in the present day, that the only relations which the mind can discover are those of agreement and difference. This is another of the ways in which sensationalists and positivists are narrowing the capacities of the human mind and undermining our belief in the reality of things. They first represent us as incapable of knowing things. Then they make the relations not to be in the things. Thirdly, they speak of agreements and differences as the only relations which the mind can discover. Having so limited human capacity, many are prepared to account for it by material agency, or simply by the action of powers unknown. But to discover that things agree or disagree we must know something of the things; we must know some of the qualities of the things. Farther, we discover more or less clearly what it is that they agree or disagree in ; it must be in form or property or something else known or conceived.

CHAPTER II.

WE see a tree in full blossom. I. We discover that this tree is the same as we saw yesterday, though the blossoms are farther advanced. II. We contemplate separately the blossoms, but as blossoms of the tree. III. We notice that the tree resembles others standing near it. IV. We observe the shape and size of the tree. V. We calculate how long the blossoms continue. VI. We try to estimate the number of blossoms. VII. We find that they emit a pleasant odor. VIII. We discover that some are blown away by the wind. We thus find that the mind of man can perceive eight kinds of relation : —

I. Identity.	V. Time.
II. Whole and Parts.	VI. Quantity.
III. Resemblance.	VII. Active Property.
IV. Space.	VIII. Cause and Effect.

I am sure that the mind can discover all these kinds of Relations.

SECTION I.

RELATION OF IDENTITY AND DIFFERENCE.

This relation carries us back to the Simple Cognitive Powers. We have seen that we know objects without and within us as having Being. (Pages 77, 78.) The same object may be presented to us at different times and it may be with different concomitants, and when we declare it to be the same the judgment is one of identity.

We have an immediate and direct means of knowing one kind of identity, and that is our personal identity. First, in every act of consciousness we know self as having Being. (Page 70.) Again, in every act of memory we have a remembrance of past self and also a consciousness of present self, and on comparing them, we at once pronounce the two to be the same. There may have been many and varied differences between the two states. In the past state remembered we may have been hopeful, elastic, joyous; in the latter, sad, depressed, gloomy; yet we discern an essential self that is the same. All this is self-evident, that is, evident on the bare contemplation of the objects. It is necessary; we cannot be made to decide otherwise, or allow for an instant that we are different persons from what we were a month or a year or ten years ago. It is also universal, that is, entertained by all men. We are thus entitled to regard it as intuitive, for it can stand the tests of intuition.

We have no such direct means of knowing the identity of other and external things. I saw a man with a white coat yesterday and I see a man with a black coat to-day. I have no intuitive means of knowing that it is the same man. I know, indeed, that everything we know has being, — the thing we remembered in the past and the thing perceived at present; but I have no intuitive means of knowing that they are the same. We are here thrown upon experience, which experience always falls back, however, upon the principle that everything has being. But it is by a gathered induction and inference that we are able to decide that this person or this table we now see is the same as that we saw yesterday. Hence, while there is no room for difference of judgment as to our personal identity, there is room for difference of opinion as to the identity of other things; and there are often mistakes and disputes as to the identity of persons. We have in these last cases to depend on the common rules of experiential evidence, taking care not to decide one way or other when we have no sufficient proof.

The Relation of Identity is always one and the same, but it may take three distinct forms; Identity Proper, Contradiction, and Excluded Middle. These were mixed up together and at times confounded in the ancient philosophy, in the mediæval ages, and indeed till the present century, when they have been distinguished and carefully separated, greatly to the benefit both of Logic and Metaphysics.

Identity Proper. — The same is the same; A is A. This proposition is apt to appear very trivial and even silly till we put it in this form: "The same is the same, observed it may be in different circumstances, or with different associations." I am sure that I am the same person to-day when I am in good humor as yesterday when I was angry. I discover that this man in robust health, florid and active, is the man I saw a year ago, pale, sallow, and listless. This same faculty guides us in all those judgments (affirmative) which are called immediate inferences or derivative judgments, — as, when it is given that all men have a conscience, we argue that the Indian has a conscience. It also regulates all cases (affirmative) in which from two premises we draw a conclusion, — as, when it is given that "he who is responsible has free will," and that " man is responsible," we infer that " man has free will." In such cases the mind discovers an identity in thought in propositions which differ from each other in form, in language, and in extent.

The Principle of Contradiction. — Here when we have a cognition or an idea of a thing, we are prepared to deny that it has not those qualities which it is regarded by us as possessing. As, knowing that this body has a square shape, we deny that it is round. As, knowing what mammals are, we deny that they are not warm-blooded. Our negations, like our affirmations, thus carry us back to our knowledge and our ideas.

The principle of Contradiction has been expressed variously; one is, A cannot be not A. The best form, I think, is the old one so much used by the mediæval logicians : " It is impossible for the same thing to be and not to be at the same time." This principle applies both to things and their qualities. If I know that stone to exist I cannot allow that it does not exist, and I must contradict those who so assert. Again, if I know that I have free will, I must deny that I have not free will. If I know that this body is extended, I put a negative on all asseverations that it is not extended.

This principle regulates all propositions which draw a negative proposition by immediate inference. Thus, it being allowed that no man is infallible, we infer that the pope and the public press are not infallible. It also rules reasoning in which the conclusion is negative. On being allowed that one who has not reason is not responsible, and that this man is without reason, we argue that he is not responsible.

Excluded Middle. — When two propositions are contradictory both cannot be true. If this man has free will it cannot be that he has not free will. When the two propositions are truly contradictory one or the other must be true. If John Smith did commit the robbery, it cannot be that he did not commit it. But it is to be observed that propositions may seem to be contradictory when they are not so, in which case they may be both true or both false. Thus, man may be free while yet causation acts in his voluntary acts. I may be able in one sense to conceive of space and time as unbounded, that is, I decide intellectually that they are so; and in another sense, I am obliged by my nature to conceive of them, that is, image them, as being bounded.

It has been shown by philosophic logicians that these three laws regulate all discursive thought. But it is to be noticed that discursive thought always implies something admitted on which it proceeds. All our immediate inferences and reasonings thus carry us back to our primitive cognitions, beliefs, and admitted judgments.

SECTION II.

RELATION OF WHOLE AND PARTS.

When we consider the relation of the whole to the parts, this is Comprehension. When we consider the relation of a part to the whole, this is Abstraction. When we separate the whole into its parts, supposed to be its whole parts, this is Analysis. When we put the parts together to make up the whole, this is Synthesis. These are operations which every one is performing every day. In their higher forms they act an important part in science of every kind.

In the ordinary affairs of life we have ever to break down the concrete or complex whole into its parts, to contemplate and use separately what we have seen together, and to combine things in order to make up a connected whole; to combine, for example, the separate sides and rooms of a house to make up our idea of the house. We are ever required to consider the attributes of things as well as the things themselves. In a loose way we are ever distributing things into compartments and putting together the compartments to make a complete conception.

Required even in practical matters, these are essential processes in every kind of scientific investigation. In nature different agencies are so mixed together that if we would ascertain their mode of operation we must separate them. Inductive science, says Bacon, begins with "the necessary rejections and exclusions," or, as Whewell expresses it, with "the decomposition of facts." Abstraction is necessary in order to our thinking of, or inquiring into, any attribute, quality, or law. Analysis must be constantly employed in every kind of investigation, physical or metaphysical. It is equally true that in all scientific inquiry we ever aim at reaching a synthesis of the things we have considered separately.

SECTION III.

RELATION OF RESEMBLANCE.

In our observation of this relation, as of every other, we proceed on our knowledge or idea, previous or present, of objects. From the knowledge or idea we have of them we perceive that there are points in which they are alike. This enables us to put them into a class, to which we may attach a name. That class must include all the objects possessing the common attributes fixed on. By the faculty which discovers whole and parts we get, as we have seen, our abstract notions. By the faculty which discovers relations of resemblance we get our general notions or concepts. These two kinds of notions are not to be confounded. By abstraction we have an idea of an attribute. In our general notions we put things together that have a common quality. From this it appears that abstraction, which fixes on the common attribute, is necessary to generalization. It has to be added that, after general notions have been formed, we can compare them and form higher and more complex concepts.

(1.) It should be noticed that in all cases generalization proceeds on common properties supposed to be in the things. The concepts are no doubt formed by the mind, but they are formed from things known or apprehended. When the things are imaginary of course the notions may also be imaginary. But when the things are real the concept has also a reality, that is, a reality in the common properties possessed by all the objects embraced in the class.

(2.) The human intellect by its native tendency is ever led to seek out points of resemblance among the objects which fall under its notice. As the singular objects pressing themselves on our attention are innumerable, and would be a burden on the memory were it obliged to carry them all, so we are compelled to form them into classes, were it only to enable us to bear them about with us more readily. While they and their properties are so numerous, they all proceed from a few elements, each with a few qualities, and so we are everywhere presented with likenesses which the mind is not slow to observe. Some of these are superficial and accidental, having no importance in the arrangement of the objects in the world, but others have a deep significance as proceeding from some universally acting cause, or falling out according to a law in which there is a Divine purpose. These furnish us with the means of arranging natural objects into what may be called Natural Classes, — into species and genera, orders and kingdoms, which have a deep meaning, and open to our view the nature of the order that pervades the universe.

(3.) It is of importance to distinguish between the relation of identity and that of likeness. In the one there is a sameness in that which constitutes the being of a thing, in the other in one or more of its qualities.

SECTION IV.

RELATIONS OF SPACE.

We can discover these because we have a knowledge of objects, say our own bodily frames and bodies in contact with them, as extended, that is, occupying space. We are now able to compare bodies in respect of the space which they occupy, and thus determine their form and size, linear, superficial, and solid. By this gift of Locality, as it may be called, we are able to estimate the distance and the bulk of objects, and to determine what they are as we meet them, say man or woman, boy or girl, horse or cow, tree or rock, river or mountain, often at a great distance. In all such cases we seem to settle on a unit of some kind, of shape or distance, and to fix on a line of direction, say in a straight line from our eye or ear, and to bring all things into a relation to these. The skillful and practised eye, or rather mind acting through the eye, may attain a wonderful accuracy, apart from the use of any instrument, in determining these special relations.

By a process of abstraction we can separate the space from the bodies in space and then discover the relations of pure space. This is what is done in geometry. We define the things we are to look at, line, surface, triangle, square, circle, and then proceed to compare the things defined. Some of the truths we discover by pure intuition, that is, by the bare contemplation of the figures. Thus on considering two parallel lines we declare that they will never meet. In other cases we cannot discover the relation directly and we resort to mediate reasoning. We have found that $A = B$ and $B = C$ and we conclude that $A = C$. We do not require any enunciated general rule to enable us to do so. We so conclude at once on the bare contemplation of the objects. But then some good purposes are served by expressing in a general form the principle on which we have proceeded, which is, " things which are equal to the same thing are equal to one another." This may now be announced as an axiom regulating our reasonings. A corollary is a truth derived at once from some truth we have demonstrated.

SECTION V.

THE RELATIONS OF TIME.

We can discover these because we already, by memory (the Recognitive Power, p. 153), have an apprehension of events as happening in time. These relations are not so numerous as those of space, but are of equal importance. They may be summed up under three heads; contemporaneous, prior, and posterior. Here, as in regard to space, we have to take a unit of comparison, a second, a minute, an hour, a year, a century, and estimate all things by it. The digits give us our decimal units and the seasons the yearly units. Great events, such as the Jewish passover, the institution of the Olympic games, the birth of Christ, the flight of Mohammed, give our starting-points in historical chronology. The fossils with the minerals give us the epochs in geology. By such means we can go far back into the past, and by reasoning from the past look far forward into the future.

SECTION VI.

RELATIONS OF QUANTITY.

These are the relations of less or more, of degree of proportion. We can discover these because we have ·had objects before us with bulk and events running through time, and also because we have discovered relations between these, such as relations of space and time, and it may be all other relations. Having discovered objects and relations, we can find that they have less or more of the qualities we have fixed on and specify the proportion between the qualities. In the practical affairs of life this capacity keeps things in their proper place, calls forth our acts at the suitable time, imparts a unity and a consistency to the conduct, and makes things march in harmony.

This is specially the mathematical talent. In geometry, indeed, the relations of space are the main ones looked at. In arithmetic we may have to use the units supplied by time. But ever since Descartes showed that the relations of space could be expressed quantitatively, mathematics, as a science, may be represented as the science of quantity and as dealing with the relations of quantity.

It is of importance to show that the relation of equality is not the same as that of identity or as that of resemblance. In the judgments of identity we declare the objects to be the same. In those of resemblance we proclaim them to have like qualities. But in equality we declare them to be the same in point of quantity. When we declare that A = 6 or that A resembles B, we do not affirm the things to be identical or that they are like, but that they are equal. By applying these distinctions we are able to correct a mistake which certain mathematicians are seeking to introduce into logic. They interpret the proposition "all men are mortal" as meaning "all men = some mortals." Now this is to misunderstand and pervert the proposition, which, when properly interpreted, means that "all men have the attribute of mortality," a proposition in comprehension (whole or parts), and the involved proposition in extension, that is, resemblance, "all men are included in the class of mortals."

SECTION VII.

RELATIONS OF ACTIVE POWER OR PROPERTY.

We are able to know these relations because we know objects without and within us as exercising power. We know body as having potency probably by all the senses : we seem to have a perception of body, affecting us even by smell, taste, feeling, hearing, and vision ; certainly we have it by the muscular sense. It is palpably wrong to assert, as some do, that body is altogether passive. There is a sense, it should be admitted, in which body is passive. If left isolated and alone it will not act ; it will continue in the state in which it is. But every body is acted on by other bodies, it is attracted by them or chemically affected by them, it acts and is acted on, it moves and is changed. All action of bodies is mutual action : one body acts on another and the other acts on it. In respect of their active powers bodies have various relations to each other which we can discover and express. Physical science consists essentially in the discovery of the relations of bodies to each other, which are expressed in laws, mechanical, chemical, vital.

It is generally acknowledged that mind has power. I think I see proof that body acts on mind and mind on body : an action of the nerves and brain gives rise to perception: I will to move my arm and it moves. There is always much mystery about the relations, that is, mutual actions, of mind and body; still some points have been determined as to the relation of mind and the cerebrospinal mass, and hundreds are eagerly employed in seeking to make farther discoveries. We certainly know much speculatively and practically as to the activity of mind and the laws which govern it. From the days of Aristotle there has been a science of mind, and it has made considerable progress in modern times. This treatise is professedly an endeavor to discover the powers of the mind and the relations between them. In the business of life and the intercourse of mankind, one man seeks to sway his neighbor by working upon what he knows of the motives by which he is swayed.

SECTION VIII. .
RELATION OF CAUSE AND EFFECT.

Causation may be considered Objectively and Subjectively. Under the former aspect we regard it as acting independently of our observation or any observation of it. A spark will kindle gunpowder whether we notice it or not. Under the latter we contemplate the mind looking at it; or, in other words, we inquire what is the nature of the exercise or power which discovers the relation.

(1.) *Causation Objective.* — Much remains to be settled as to what causation is. How does force stand related to cause ? How are properties related to cause, when it is said that mind and matter are known by their properties ? What is the difference between power in mind and power in matter ? Some points seem to me to be determined, and these may in the end determine the others.

First, there is an energy in all physical nature. It is acknowledged that there is a Conservation of Energy — Spencer calls it Persistence of Force. Physicists distinguish between Potential and Real Energy — (Aristotle's distinction between δύναμις and ἐνέργεια.) The former cannot be increased or diminished by any mundane agency, — by any power but that of God to whom in the end all power belongeth. All the physical forces, mechanical, chemical, electric, magnetic,— some think the vital also, — are correlated and can be transmuted into one another, so much chemical and electric power being an equivalent of so much mechanical energy. This power is always in body, but may be transferred from body to body according to the capacities of the body. Thus a ball A in motion strikes a ball B at rest, and the power in A is transferred to B, which moves while A now rests. It should be observed that while the amount of energy in each body has changed, the whole amount of energy continues the same. I believe the capacity for energy in the body also continues the same, and it is possible to reverse the action and make B in motion strike A at rest and transfer its motion to it. Every body has a certain capacity (δύναμις) for receiving this power ; and this power, in exercise, constitutes the properties of the body ; its gravitating, chemical, electric, magnetic, and, it may be, vital, properties.

As to mental properties, — say intelligence, emotion, moral approbation, will, — there is no reason to believe that they are correlated with physical powers. It is the office of psychology to determine their nature, their extent, and their limits. In doing this it labors under the disadvantage of not having a precise standard of measurement as mechanics have in foot-pounds ; but it has a counteracting advantage in the acts being under the immediate cognizance of the consciousness.

Secondly, another important point has been established. John S. Mill has shown that there are always two or more agents in a cause (physical). We are accustomed to say that this plant was killed by the frost. But there is more embraced in the cause than the frost, that is, than the low state of the atmosphere; that agency alone would not have produced the effect. In the cause we have to include the state of the plant. The frost might not have destroyed the plant if it had not been tender. The low temperature and the tenderness of the plant together constitute the cause and were necessary to cause the effect. Carrying out the same views a step farther, I have been endeavoring to show that not only is there a duality or plurality in the cause, there is the same in the effect. There is a change in the plant, but there is also a change produced, difficult to measure, in the temperature of the atmosphere. It is the same in all cases, two or more agents acting as the cause and the same agents changed in the effect. A ball A strikes a ball B, both balls act, and both balls are acted on and are changed, the one losing momentum, the other gaining it. There is thus a most complicated agency in causation and a like complication in effectuation. How numerous the agencies producing any given historical event! On the supposition that the wolf suckled Romulus, we may trace the influence on the whole history of the Roman people. Certainly the character of Knox has so far influenced the Scottish character in all later ages. The Puritan character of the seventeenth century and the Pilgrim Fathers, modified by very different influences, has helped to mould the people of New England.

In the common explanations one of the agents, the more prominent one, or that supposed to be the main one, is spoken of as the cause. The others are described as the conditions or the occasions — it was on the occasion of this man being exposed to cold that he caught fever and died. We commonly look only to one part of the complex consequence, the one most prominent or the one expected, as the effect, and we call the others incidents or concomitants. We say that the flood refreshed the ground, but incidentally or accidentally it also drowned a certain person. But, rigidly speaking, there is no chance in what is occurring, no accident in what has taken place. All the agents acting are to be included in the cause, and are also to be seen in the effect. They constitute the invariable antecedent which has produced the effect, and which when it recurs will forever produce the effect. The same effect precisely will follow when they are all present.

Cause and effect do not consist, as Hume maintains, in invariable antecedence and consequence. In the cause, that is, in the agents forming the cause, there is power, force, or energy to produce the effect. It is not the invariable antecedence which makes the cause, but the cause which makes the invariable antecedence.

All power, we have seen, resides in a substance (p. 79). Let us suppose, as we have done (p. 223), that every substance is endowed with its own capacity or property; then, the substances continuing the same, there must always be the same amount of energy in the world, — a conservation of energy, a persistence of force, as it is called. The grand doctrine of our day, of the conservation of energy, seems to follow from these principles.

Causation Subjective. — This falls under Psychology. This relation, like every other, throws us back on our primitive cognitions. We know mind by self-consciousness and body by sense-perception, as possessing power to produce an effect. Our earliest knowledge of mind

15

and our earliest knowledge of matter is thus associated with efficiency. Herbert Spencer may be right in representing force as the most essential quality of body as made known to us. It is certainly known as early and directly as extension, commonly regarded, and I believe justly, as one of the essential qualities of body. It may be by resistance, that is, force from a surface, that extension is first made known to us by the touch and by the rods in the eye. Power is more fully revealed to us in the exercise of mental properties and we regard it as an essential quality of mind.

We trace everything that occurs to a power in a substance producing it. This is a primitive perception. It is self-evident, evident in the thing itself as we know it. It is necessary; we cannot be made to think or believe otherwise. It is a universal perception. Children act upon the conviction as soon as they begin to act intelligently; they follow the light which they find produces the pleasant impression on their eye. Savages, even the lowest in the scale, act upon it, and expect the same effect to follow the same cause. Not that they are able, like a metaphysician, to enunciate the law, but upon the bare inspection of the object before them, they form a decision and act upon it.

It is after careful introspection and reflection that we are able to detect the precise nature of the law and to formulate it. The law is not, as most people, who have not thought much on the subject, are disposed to say, that everything has a cause. If this were the law, there would be no first cause, and we would require to seek for a cause of God himself. Our primary knowledge of power is of a new thing produced. We instinctively seek for a cause only for a new object or a new manifestation of an old object. The true expression of the law is that " whatever begins to be has a cause." This is to our minds a fundamental law at the basis of all action.

While our primary conviction as to cause and effect is intuitive, yet much of the knowledge which we have of actual causes and effects is the result of a gathered induction. It is only by careful observation that we know the nature of particular powers, such as gravitation, chemical affinity, electricity. It is by careful weighing and measuring that we know what are the powers in any one bodily object, say what is its weight, or its chemical affinity towards any other body. But believing that every material object has power we are prompted to find what are the extent and the limits of that power. We see that intuitive conviction, so far from restricting experimental investigation or rendering it unnecessary, is the main means of inducing us to engage in it ; for while it does not reveal the cause, it constrains us to believe that there is a cause, which we therefore inquire after.

It should be noticed that the belief in the relation of cause and effect is not the same as the belief in the uniformity of nature. These two have often been confounded. Though connected, they are essentially different. The former is intuitive and universal, the latter is a discovery of science and is not universally believed in. The child and the savage always look for a cause to every phenomenon in which they are interested. But they have no special faith in the uniformity of nature. Till lengthened observation — till, in fact, advanced science — teaches them, they are quite ready to believe that nature, so far from being under law, is acted on by various supernatural agencies, and is under agencies always acting causally, but in no rigid order. It is only as people advance in knowledge that they discover that all events obey laws narrower or wider. It is only within the last few ages that the uniformity of nature has been established as a scientific truth.

But we have here to do not with the uniformity of nature, but with causation as a law of mind. According to the account given above causation is always in objects, material or mental, all of which possess power. As there is reality in objects, material and mental, there is reality in the powers and in their causal relation by means of these powers. The cause of a known effect is not super-induced upon the objects by the mind (as Kant holds); it is perceived as in the objects and in the nature of the objects. We are thus in a real world not only in regard to objects, but in regard to all their action, which is indeed an essential part of their nature. These coal strata which we see in the earth are a reality, and we argue from them that the deposited plants which formed them millions of years ago are also realities. By a like rea-soning process, as we discover these adaptations in the eye and ear so wonderful, we seek for a cause which is also real in the designing mind of the living and true God.

But while all this is true, there may be limitations to the truth. We know power to be both in mind and body and in their very nature. But it does not follow that every act is one of causation and necessary causa-tion. As a matter of fact we have a peculiar conscious-ness as to acts of the will when we choose this, and reject that; when, for example, we resist the evil and choose the good. Because we believe in a cause be-hind every other action of body and mind I am not sure that we are required to seek for a causal power behind these free acts of the will, or at least that this causal power is of the same kind as operates so sternly in every other part of nature. There may be an outlet here for free will in perfect consistency with the universal prevalence of causation in all other parts of nature, including all other mental states, intellectual and emotional. I do not claim that in this way we can clear up all the mystery which broods like a cloud over the point at which free will and causation meet. But we have shown that there is a place where free will may act in perfect consistency with all that we know of causation ; thus allowing our consciousness to give its testimony in favor of free will without interfering with the dicta of any other part of our nature. The subject will be resumed when we come to speak of the Will in a subsequent volume.

CHAPTER III.

By these we proceed from something given or allowed to something else derived from it by the simple exercises of thought directed to the objects. They are commonly represented as being Simple Apprehension, Judgment, and Reasoning. These are all performed by the Comparative Powers, specially by three of them: the faculties which discover the relations of Identity, Comprehension, and Resemblance.

(1.) *Simple Apprehension*, the product of which is the Notion. — There are three kinds of Notions: the Singular, the Abstract, and the General (Concept). The Singular Notion is given us originally by the Simple Cognitive Faculties of Sense-Perception and Self-Consciousness. Upon this we may perform discursive processes and still keep it singular. Thus "Socrates" is a singular term, which we are enabled to apprehend because we know ourselves by the two original inlets of knowledge. "This man" is also a singular term, though we have performed an intellectual process and referred the individual to the class Man. In the Singular Notion there is no exercise of the Comparative Powers. It comes into Logic simply among the things given or allowed, and not among the processes. The Abstract Notion is formed by what we may call the power of Comprehension; it is the notion of an attribute. The General

Notion or Concept is the product of the faculty of Resemblance; it is the Notion of objects joined by their possessing common attributes.

(2.) *Judgment.* — In this we compare Notions with the view of declaring their agreement or disagreement in a proposition, affirmative or negative. Our judgments proceed on our notions, and our singular notions carry us back to our primitive cognitions and beliefs, and our abstract and general notions imply previous acts of comparison, involving previous cognitions or ideas. Our judgments passed on notions have thus a reference to things or imaginations formed out of things. Our judgments may be of three kinds. They may declare an Identity, — as when we say, " Metaphysics is the science of First Principles ; " " Logic is the science of the Laws of Discursive Thought." Or they may be judgments of Comprehension, — as when we say, "The dog barks," where we make barking an attribute of the dog. Or it may be one of Extension, that is, of Classes or General Notions ; thus we may interpret the last example as meaning " dogs are in the class of barking animals."

(3.) *Reasoning.* — It is acknowledged that this is a form of Judgment in which we have three notions instead of two, and compare two notions by means of a third. " The New Zealander, as he has the power of speech, is a man." Here we compare " New Zealander" and " man " by means of possessing the power of speech. We have already seen that the principle of identity regulates many of our ratiocinations. (See pp. 213–215.) So far as reasoning in Extension — the reasoning treated in the common logical treatises — is concerned the principle of resemblance is involved — the resemblance of the objects in the concepts. " This horse being a mammal is warm-blooded." Here we place " horse " in the class

Mammal, and therefore in the class of warm-blooded animals. There may also be reasoning in Comprehension, in which we look to the attribute, — as when we say "this man, having intelligence, conscience, and free will, is responsible;" where it is argued that, the attributes of intelligence, conscience and freedom involving responsibility, man as possessing these must be responsible. Reasoning in Comprehension may always be translated into reasoning in Extension.

Logic does not, every one now acknowledges, give us the power of reasoning or discursive thought; it simply expounds the process involved. We think and reason spontaneously, then reflect upon what has passed in our minds, and may express the operation in formal laws. It follows that if we have given the proper account of the logical laws we have unfolded the laws of our ordinary processes of thinking from day to day in the common affairs of life. Every man is exercising continually the faculties which have just passed under our notice, and what psychology does is to unfold the nature of these faculties; while it is the function of logic to formulate them into laws by which we may test discursive thought, may justify the truth and expose the error.

CHAPTER IV.

INTUITION IN THE DISCOVERY OF RELATIONS.

I HAVE been showing in this work that there is intuition, that is, the immediate perception of objects, both by the senses and self-consciousness. According to the school of Kant there is no other intuition than that of sense, external and internal, and all beyond is subjective and formed by the mind. It is a curious circumstance that, according to Locke, there is no intuition of sense, there is only one of judgment; that is, the perception of the immediate agreement or disagreement of our ideas. (Essay, B. IV. c. 11.) I regard both these views as so far erroneous.

We have intuition of body without and self within. But this is not, as Kant holds, of mere phenomena in the sense of appearances, but of things; of body as extended and resisting energy and of self as thinking or feeling in some particular mode. But there is also a sense, and an important one, in which we have also an intuition of relations. We first perceive things and then also the relation of things, and some of these may be known by intuition. We know matter as existing, but we also know, and this directly, that it has relation to other things known, that it is in space, and that there is causation in its action. We also know mind as existing, and we know it to have being, potency, spirituality, thinking, and relations to things.

Most important consequences follow. We not only know things, such as body and mind, but things perceived in them and in relation to them, to be realities; and both alike realities. We know mind as having extension, and we know mind as thinking, say as contemplating extension, and we know the one as well as the other with immediate certainty. I hold that as the things are real so the relations in the things are also real. In holding this doctrine we save ourselves at once from the idealism of Locke and the *a priori* forms of Kant. They are in error who hold that all knowledge is relative, that is, only of the relations of things themselves unknown; and they are equally in error who affirm that relations are forms added to things by the mind. The relations are in the things, and are as real as the things, only with a somewhat different kind of reality, a sort of dependent reality in the things. True, we only know individual things by the senses, but we know by contemplating them that they have relations. In this way we reach a realism according to which the mind knows things and their connections.

CHAPTER V.

I. THESE Faculties are in all men; not merely in certain individuals, times, or nations, but the properties of humanity. They are found in a rudimentary condition in children and in idiots, in the former to be developed. Madmen often display them in an intense form.

II. They constitute the highest of the intellectual powers; I may show that the moral are higher. They carry us out the farthest and they raise us up the highest. They enable us to connect all things we know with one another, and they take us as far out as the connections reach. Thus causation takes us as far back as the millions of geological ages, and as far forward as the causes now in operation go, — show us, for instance, that this world is to be burned with fire.

III. In their exercise they have risen very much above the senses, and the need of the coöperation of the senses, and of the bodily frame generally. True, they have been dependent on these for the materials on which they have to pronounce a judgment, but the judgments themselves are purely mental. Hence we often find that in old age, when the senses, the memory, and the informing faculties are breaking down, the judgment is as sound as ever and fails only when a proper statement of facts is not given it.

IV. They have all a tendency to operate and seek out

for the appropriate objects, allured by the numerous relations which may be discovered in them. Thus we have pleasure in finding an essential sameness in the midst of minor diversities. We love to visit a locality with which we were familiar in former days, and to trace the identity in the hills and valleys, so changed in the houses upon them and the people dwelling in them. We are interested in the lights and shadows of the landscapes and the varying aspects of the sea and sky. We set ourselves keenly to detect an old friend whom age has changed. We analyze the bodies in nature and seek to solve the difficult problems in science and philosophy. We love to resolve a complex whole into its component parts, and to understand thereby the whole of which they form a part; and we feel as if we know a thing only when we are acquainted with its constituents. We delight to trace the likenesses among objects, and to discover the analogies between things often far removed from each other, and which bind in a unity all parts of nature and of history. We find it pleasant, as well as profitable, to observe how plants and animals are after a type; how the heavenly bodies move in like curves, elliptical or spiral; and how occurrences, historical and cosmic, move on in epochs. The idea was anticipated by Pythagoras, and has been established in modern times, that physical laws, such as gravity and chemical affinity, take a quantitative expression. We like to see activity in the breeze, in the running stream, in the leaping cataract, in the rippled ocean; in the perpetual motion and prattle of boys and girls, in the contests of wit, and the Demosthenic torrent of eloquence. All lofty minds delight to follow effect to cause, and cause to prior cause, on to the great originating Cause from whom all things proceed.

V. Our comparative faculties are admirably suited to the state of things in which we are placed. I can conceive of a world in which there is no such adaptation; in which the relations which we are inclined to notice have no correspondences in what falls under our eye. There might thus have been an inscription without the means of deciphering it, or a writing without an interpretation. But we find instead that we live in a world in which there is a beautiful harmony between the eye that looks and the forms and colors which it gazes on. We feel security in falling back with the Eleatics from the phenomenal variations revealed by the senses, upon the permanent τὸ ὄν, revealed to the intellect as a νοούμενον, and having a correspondence in the permanent mind and the conservation of energy in matter. Our analytic propensity is rewarded in discovering that complexities can be resolved into their elements. Our inclination to generalize is encouraged by finding that organic objects can be arranged into species, orders, and kingdoms. We rejoice when in accordance with our anticipations we find all nature conformed to laws of time, figure, and proportion. We are moved by the movements of nature in heaven above and the earth beneath, in the rapidity of molecules and the quickness of thought. We pursue eagerly one act to a previous one, and are stayed only when we rest on one unchangeable substance. The consequence of all this is that instead of being strangers, wanderers, or outcasts in the world in which we are placed, we are, as it were, among friends, with a Friend who is the bond of union among them all.

VI. There is a correspondence between the subjective and objective worlds, between the thinking mind and the objects it is called to think on. This does not arise, as the ancient Eleatics and modern pantheists maintain,

from the unity of thought and being; for we have as
clear proof of the difference of mind and matter as we
have of their connection. Nor can it spring solely or
even mainly from the two having acted on each other
for indefinite ages and become adjusted, — as Herbert
Spencer accounts for their relation. No doubt their
connection may have influenced both: thus, the contemplation of the action of matter by mind may have created tendencies in mind; but to produce this fruit there
must have been an original marriage union. In another department of nature we are prepared to acknowledge that the rays of light have not produced the eye,
nor the eye the rays of light, though they have so far
brought each other into conformity; so neither does the
subjective mind create objective matter, nor objective
matter create the properties of mind. We are thus
driven to the conclusion that there must have been a
foreordained conformity between them. We have thus
the true doctrine of preëstablished harmony between
mind and body, of which Leibnitz had after all only imperfect glimpses; not a harmony of the two acting apart,
but of the one acting upon and with the other.

VII. We have seen that there is an intimate connection between
our associations and the discovery of relations, and have illustrated
this by Resemblance and Contrast (pp. 147, 148). But the remark
holds true of all relations. For every relation discovered there is a
ground, and this may become the bond of an association which strengthens and enlarges that of contiguity. (1.) On seeing a man in one dress
to-day we think of him in the other dress in which we saw him yesterday, the man himself being the same in both. (2.) On the leg of a
table being seen by us the idea of the whole table is apt to come up.
(3.) We have already discussed resemblance. (4.) Certain relations
of triangles suggest other relations to the mathematician, also of triangles. Stratford and Shakespeare suggest each other, because of the
birth there of the great poet. (5.) The year 1790 is apt to bring up
both the French and American revolutions, in both there being a rev.

olution about the same time. (6.) The proportions of one figure recall those of another. (7.) The activity of some one thing, such as life, calls up the activity of other things, such as the wind. (8.) The view we have given of physical causation, that it consists of two or more agents in the cause, to be found in a changed state in the effect, enables us to see how the common qualities of the one should call up the other. In all cases, the powers in the substance acting in the effect suggest the powers acting in the cause, it may be with their adjuncts, and *vice versa.*

Associations in all cases imply a contiguity, but in the highest forms correlative Associations are strengthened and enlarged by the discovery of relations. I may add that a man's intellectual wealth is large in proportion to the formed coins and cut diamonds which he has laid up in correlations. Without many and varied connections there can be no readiness or comprehensiveness of memory, such as is to be found in our greater men. With such accumulated riches a man is ready to expend bounty of thought wherever he goes.

VIII. The Comparative Faculties differ widely in the case of different individuals. This may arise from the intensity of the original cognition, or the strength of the comparative faculty; from one or from both. In some cases it looks as if it were the original impression, — say of form, or color, or incident, which is so keen that it penetrates into us. In other cases it looks as if it were a strong intellectual energy seeking for relations.

It is evident that there are native tastes and talents. The two commonly go together, the taste calling the talent into exercise, and the talent forming and evoking the taste, and both seeking out fitting objects. When these are very marked they commonly determine the decision of the youth as to his pursuits, — his vocation, his business, his profession, his literary or scientific studies. It is true that circumstances often have a swaying influence — in fact, compel a settlement. But in most youths of any force of character there is a natural ability or inclination which selects his life for him, and this frequently

not in concurrence with outward position, but in opposition to it. If a man, for instance, has a taste for some
particular pursuit, he will be found pursuing it in his
vacant hours when obliged to engage habitually in far
different work. How often is the merchant or lawyer
longing for a leisure day or week to enable him to examine the forms of plants, and rushing forth whenever the
pressure of business allows into the midst of the beauties
of nature? How often does the minister of religion, busy
for most of the week in caring for his flock, find an idle
day or evening in which to pursue philosophic specula-
tion?

It is thus clear that one man may have a strong tendency to observe one kind of relation and another a different kind. It is always to be remembered, however, that
the same natural talent may be exercised on different objects, and it is here that external circumstances may have
a modifying influence. It may be mere accident which
determines a man with a certain taste to betake himself
to the study of plants or animals, or to painting or sculp-
ture. It is also to be borne in mind that some pursuits
require the exercise of more than one faculty, and it is
only when there is the necessary combination that the
qualification for the particular work is secured. But
making the needful allowances, it will be found that, while
in a few there is a universal ability, and in the great
body of mankind there is a moderate degree of various
talents, in some there are peculiar gifts which will and
ought to find congenial pursuits and thus determine their
destiny. It is a most happy thing when a youth comes
to know what are his peculiar qualifications, and is enabled to put them to proper use. It is a blessed thing
when a man with marked endowments is led to conse-
crate them to a high end.

IX. All the relations enumerated, with one exception, are to be found in mental as well as in material objects. That one is the relation of Space ; we cannot correlate by it our ideas and emotions. We arrange bodies according to type, but we cannot thus distribute or even conceive of our mental states. The reason is obvious ; consciousness as a faculty of intuitive cognition does not make known to us our mental states as occupying space, and as extension does not appear in the primitive knowledge, so it does not reappear in the discovery of relations. This might be urged as an argument of some force in favor of the immateriality of the soul ; we cannot conceive mental as we do physical facts, under spatial relations.

X. I need not dwell, at the close, on the means of training the powers, for I' have been illustrating them throughout. It is enough to exhort every young man to let none of these intellectual faculties lie dormant, or yield to the temptation to satisfy himself with sensations, feelings, and passions.

CONCLUSION.

RISE OF OUR IDEAS.

WE have traced the powers of intelligence from the lowest to the highest, and have shown how our cognitions and ideas arise. From every separate faculty, as they have been arranged, we get one or more of these.

We receive knowledge, probably our primary, from the senses. We thus come to know body and its modifications, especially its essential qualities, Extension and Resisting Force. We thus get our idea of Space.

A large school, the Sensationalists, maintain that we get all our ideas from Sensation. This is a fundamental mistake. We have other and higher sources of knowledge.

We get cognitions and ideas from Self-Consciousness ; the knowledge of Self in its many and varied modes as I have been endeavoring to unfold them, as knowledge, conscience, feeling, and will. It is true that a full and distinct knowledge of Self, of the Ego, is a late acquisition, but from birth there is a knowledge of self in all our acts.

Locke held that we get all our ideas from Sensation and Reflection. This is likewise a mistake. From these two we get our ideas of Existing Things, bodily and spiritual. But in the exercise of the powers as we contemplate things, we get other ideas ; such as the idea of Time, when we reflect on the past, and the Infinite, as we go out in thought and conceive more and more, and yet are sure that we have not come to the end, and that what we thus believe in is Perfect, and nothing can be added to it. In these exercises, Faith is at work from the beginning, and we have a conviction of the reality of things not perceived by the senses, external or internal.

By Comparison we discover the Relations of Things, discover a universal interdependence, and extend our knowledge, indefinitely, upward, and downward, and all around, and still are among realities.

When we come to speak of the Motive Powers, we may show that we get other ideas from them : as Good and Evil, and Obligation (the Imperative) from Conscience, the Lovely and Unlovely from Emotions, and Choice and Freedom from the Will.

The scattered rays may combine to form the pure white light, that is, the idea of the all powerful and good God.

Rise of Ideas in the Minds of Children. — A number of able men in various countries are engaged in this investigation, and have given us some interesting results — only they may find some deeper ideas in the mind than they have yet brought out to view. It is evident that, while some acts of new-born babes are simply reflex or instinctive, — the result of heredity, — the cognitive powers begin to work from the time of birth, if they do not, as I think, work before.

We give some statements from Darwin's "Biographical Sketch of an Infant." The infant's eyes were fixed on a candle as early as the ninth day. Long before he was forty days old he could move his hands to his own mouth. When nearly four months old, and perhaps much earlier, from the manner in which the blood rushed into his face, it was evident that he easily got into a violent passion; when forty-five days old he was observed to smile. After more than a year he spontaneously exhibited affection by kissing his nurse. At the age of six months and eleven days he showed sympathy by his melancholy face when his nurse pretended to cry. When four and one half months old he smiled at his image in a mirror, and in two months more knew what the mirror was. Before he was a year old he understood intonations and gestures, as well as several words and short sentences. F. H. Champneys ("Mind," Vol. VI.) speaks of an infant. His eyes were fixed on a candle when he was a week old. On the fourteenth day he took notice of persons and moving objects. Smiling was reported at five and one half weeks; tears, two days before the end of the fourteenth week. Professor Stanly Hall is engaged in important inquiries as to the knowledge and ignorance of children at school age.

Some ideas, it is evident, cannot rise till there is a gathering of experience, and until relations have been discovered between things. Comparison cannot work till there are things known to compare. But within a short time after birth the intuitive principle of cause and effect seems to work, and the infant anticipates the return of the pleasant light that so attracted it. There can be little associative power exercised till ideas have come up according to the associations of contiguity and correlation; and infants, in consequence, have little control over their trains of thought. Children have no knowledge, originally, of distance, but come to grasp at objects in less than a year from their birth. A pleasure from the perception of beauty of colors appears in children as soon as they apprehend objects, and gradually rises to higher forms. In childhood, and onward, as soon as objects are apprehended, there is a seeking for connections. Will, as a natural gift, operates like intellect, from birth, but at first there is little knowledge of objects on which to work. The use of signs in thinking is found in children long before they are a year old, as, for instance, the sound of a bell announcing that supper is ready. I believe that even the mature man cannot form an adequate idea of infinity, but I agree with such profound thinkers as Anselm, Descartes, and Leibnitz, in thinking that all have it in the germ. I think I have perceived it budding in children of two years, — the mind is ever stretching beyond the now and the present. I remember distinctly of musing, when under twelve years, on the mysteries involved, and losing myself not unpleasantly in the contemplation. Moral perceptions cannot appear till there is a knowledge of other beings to whom we owe duties; but conscience and affection come forth, of course in a low form, before the intelligent child has completed his first year. The dim idea of a power beyond the visible appears before school age, and may be exalted by teaching to a belief in gods or in one God.

It may have been observed that throughout this work it has been loosely applied that at the basis of the operation of all the faculties there are fundamental truths. The tests of these can be given. They are, first, Self-Evidence, or evidence seen at once in the very thing. Secondly, as following from the first, there is Necessity; we cannot be made to think or believe otherwise. Thirdly, and confirmatory of the other two, there is Catholicity, or universal consent. The exposition of these carries us beyond Inductive Psychology into Metaphysics, or the Science of First Principles.[1]

[1] These principles are unfolded in my work on THE INTUITIONS OF THE MIND INDUCTIVELY INVESTIGATED.

www.ingramcontent.com/pod-product-compliance
Lightning Source LLC
Chambersburg PA
CBHW030759020726
47499CB00006B/1684